Conversations with Trees

Conversations with Trees

AN INTIMATE ECOLOGY

Stephanie Kaza

Illustrations by Davis Te Selle

SHAMBHALA
Boulder · 2019

Shambhala Publications, Inc.
4720 Walnut Street
Boulder, Colorado 80301
www.shambhala.com

First published in 1993 as *The Attentive Heart.*

9 8 7 6 5 4 3 2 1

Printed in the United States of America

♾This edition is printed on acid-free paper that meets the American National Standards Institute Z39.48 Standard.
♻ This book is printed on 30% postconsumer recycled paper.
For more information please visit www.shambhala.com.

Shambhala Publications is distributed worldwide by Penguin Random House, Inc., and its subsidiaries.

Designed by Steve Dyer

LIBRARY OF CONGRESS CATALOGING-IN-PUBLICATION DATA

Names: Kaza, Stephanie, author.
Title: Conversations with trees: an intimate ecology/Stephanie Kaza; illustrations by Davis Te Selle. Other titles: Attentive heart
Description: Boulder: Shambhala, 2019. | Originally published: New York: Fawcett Columbine, 1993. | Includes bibliographical references.
Identifiers: LCCN 2018050410 | ISBN 9781611806779
Subjects: LCSH: Trees. | Trees—Religious aspects—Zen Buddhism. | Trees—Social aspects. | Trees—Pacific States.
Classification: LCC QK477.K37 2018 | DDC 582.16—dc23
LC record available at https://lccn.loc.gov/2018050410

Dedicated to the trees—

May they continue to thrive and flourish on this earth,
filling our hearts with joy and inspiration.

Contents

Preface ix
Artist's Note xv
Acknowledgments xvii

Introduction 1

Part One: Reaching Out

1. Close to Water 15
2. Called by Alders 23
3. Red Fir Encounter 29
4. Friends of the Family 37
5. Maple Ecstasy 45

Part Two: Tracing the Stories

6. Reference Point 55
7. Magnetic Presence 61
8. Lifetime Lovers 69
9. Mystery Pine 77

10. The Golden Time 85
11. A Way of Looking 91

Part Three: Entering the Tangle

12. Fallen Tree 103
13. House of Wood 111
14. Bones in the Land 117
15. Overtures of Peace 129
16. Lineage of Fear 137

Part Four: Finding a Way

17. Pilgrimage 151
18. The Attentive Heart 159
19. Cutting Wood 169
20. Held by a Living Being 177
21. Gift beyond Measure 187
22. Offering of Darkness 199

Part Five: Choosing to Act

23. Arbor Day 209
24. Grand Dragon Oak 217
25. Traces of a Lifetime 227
26. Wind, Rock, and Ice 235
27. A Multitude of Voices 243

Notes 253

Preface

IN THE EARLY 1990S, WHEN I FIRST WROTE THESE essays, I was pulled by a powerful call from the trees to listen, to hear their stories, to be in their presence. I didn't know where that call would take me, but I trusted my intuition to follow. I was led to many ancient and grand beings who offered unforgettable experiences of communion. I also had to face difficult questions and troubling concerns that arose in the company of trees. Since that time, a number of the trees I wrote about have perished in dramatic but completely natural ways. The Grand Dragon Oak lost a major trunk and crashed in pieces into the canyon below. The big madrone's companion tan oak at Cloud Mountain dropped its crown in a fierce storm and collapsed to disease and weathering. The Monterey pines at Muir Beach suffered attacks by pine bark beetles, and many had to be removed as neighborhood hazards. Most recently, the unusually large manzanitas at Pepperwood

Ranch went up in a hellfire of flame during the Tubbs fire of 2018.

These individual stories reflect changing conditions for trees and forests throughout the west coast of North America. In the last few decades, droughts have become more severe and drawn out, snapping even sturdy trees under water stress. Forest fires are burning hotter and more erratically, jumping canyons and fire lines faster than they can be contained. Bark beetle infestations in the Sierra Nevada have claimed the lives of millions of ponderosa, lodgepole, Jeffrey, and sugar pines, turning mountainsides into ghost lands. Urban forests in such iconic tree cities as Portland, Oregon, are losing ground to rapid development, creating heat islands from reduced shade. All of these impacts are tied to climate change and the steady warming of the planet. It is true that forests have changed radically under earlier periods of climate change. But today's levels of greenhouse gases are the highest in forest history, and this round of climate change may prove to be shocking beyond our predictions.

In the midst of all this, it is heartening that so much more is known now about the hidden life of trees. There are more books and blogs on trees, more memoirs and novels, more nature writing and personal reflection, and thus more readers engaged with tree stories. The reports of tree communication via symbiotic fungi and complex root networks have given scientific support to something many people have sensed for a long time: Trees talk to each other!

They share information about soil and weather conditions; they alert each other to beetle attacks. They function as communities more than as individuals. All this seems astonishing, but only because it breaks through our mainstream view of trees as objects for human use.

Since the first edition of these tree stories was published, I spent twenty-four years teaching at the University of Vermont as a professor of environmental studies. There I developed my scholarship and writing on Buddhist environmental thought and further engaged the natural world from a contemplative perspective. Though I learned the East Coast trees and delighted in the maple and hemlock forests of Vermont, I often longed for my West Coast trees, with whom I felt so much more at home. When it came time to complete my university service, I again felt the powerful call from the trees. We headed west to settle in Portland, Oregon, among the towering Douglas firs of Cascadia. Now my bookshelves are lined with field guides of the Pacific Northwest, and I am learning how volcanic action, glaciation, ice age floods, fire, and climate have shaped these landscapes. I am back with my tree tribes and deeply grateful for their company once again.

And not a moment too soon, it seems. With climate change accelerating and national politics precarious, the future of the planet is at stake in a frightening way. Global warming has forced feedback systems in soils and forests into overdrive, and pollution from atmospheric carbon continues to rise. As forest cover shrinks due to agriculture, timber cutting, and urban development, forest species are disappearing all over the world. All the best efforts in conservation biology and ecological restoration have not kept up with the economic and social pressures on the forests. Many people speak of a dark despair, sensing the planet is under unacceptable levels of assault. Not only forests are suffering, but we too, as human populations, are meeting great suffering from disease, toxins, overheating, lack of water, and conflicts over resources.

In this context, what can a book such as this offer in today's world? How can a book make a difference against the relentless destructive forces that seem to be taking the planet into a dark time? I see this writing as a call for reflection, for clear seeing, an invitation to intimacy. Being with trees is a return to soul, an offering of refuge. These essays provide a place to recover from the sense of anguish and discouragement that comes in waves and erodes resilience. In my own investigations with trees, I seek a deeper truth of relationship, of reunion; I am looking for a better future for human-tree relations. I know it is possible. I am not alone in my seeking.

The themes in this book are still vital and important today: the role of stories and storytelling, the place of trees as wisdom keepers, the confrontation with the killing mind. It still makes sense to go on pilgrimage to trees with open hands and heart. I still carry my questions to trees, and I want to know—Bristlecone pines, how are you enduring? Fire-scarred bays, how will you recover? Beetle-eaten whitebarks, what will it take for you to survive? There are so many stories to be uncovered—of loss and destruction, of tree planting and recovery, of communities falling in love with their trees. When I first shared these tree stories in public, many people came up to share their own tree stories with me, looking for affirmation that they were not crazy for what they had experienced, for what they knew in their bodies and hearts. Now, twenty-five years later, the times are so much more receptive to tree stories. This is a really good thing; we need all the tree stories and tree friends we can find. These wisdom keepers have quite a lot still left to share with us newcomers on the planet.

I offer these stories as part of a wave of openheartedness for trees that is rising up right now in a beautiful affirmation of life. May they be part of the much-needed healing on the land between people and trees, and may they inspire yet more seeking and pilgrimage in the company of the old ones.

Artist's Note

THE ILLUSTRATIONS ARE REPRODUCTIONS OF HAND-printed lithographs. In this traditional process the drawings are executed directly on the textured surface of beautiful limestone slabs with a crayonlike pencil. I feel very much at home with this nineteenth-century graphic art. There is for me a strong connection with the landscape artists of that time whose work conveys a spiritual force with careful and loving presentations of the natural world.

Undertaking the work of this book has given me a way to integrate my somewhat anachronistic graphic tastes with my current environmental concerns. It has allowed me to reunite with the wandering curiosity of my boyhood rambles in the hills and creeks of northern California. These works are gestures on behalf of the misused and obliterated places that were once my solace and inspiration. These places are the very source of my desire to participate in the protection of what is yet in our care.

—DAVIS TE SELLE

Acknowledgments

MANY TREES, PEOPLE, AND PLACES HAVE MADE THIS work possible and supported it with loving care and presence. May they receive my gratitude for all they have given me.

I thank, first, the trees for the inspiration of their lives. I am especially grateful to coast live oaks and redwoods, who provided steady company and spiritual teaching during much of the writing of the book. Many other trees who do not appear in this book have also contributed much to my ongoing conversation with trees. I acknowledge the support of the beautiful oaks at Pepperwood, the lovely maples of Vermont, the trees of my night walks, and the many tropical and temperate forests who produce the oxygen I breathe.

With some pangs I offer thanks to the trees that gave their lives to become wood fires warming me, tables for writing, homes sheltering me, reference books, and all the many sheets of paper used to compile this book. Journals, sketchbooks and newsprint, pencils and art paper, several drafts of the manuscript, the bound book—all these have required the sacrifice of trees, not without some question on my part.

I am most indebted to my teachers for their kindness and dedication to the truth. To Kobun Chino Otogawa I offer my thanks for root understanding and the gift of the Dharma. To Robert Kimball I offer thanks for the power of the green circuit and for building my confidence in the early writing. To Thich Nhat Hanh I offer thanks for the practice of the precepts that guide this work. To Joanna Macy and Robert Aitken my thanks for guidance and inspiration in developing my capacity for compassion, social action work, and the wisdom of Prajña Paramita. May I honor these teachers and their gifts by the contribution of this effort.

Many intimate tree discussions and activities have taken place in the company of deep friends. I am grateful to Greg de Nevers for transmitting his love of plants to me and continuing my ongoing botany lessons. I am grateful to Jane Rogers for her steady support and shared joy in the Sierra trees. I am deeply indebted to Wendy Johnson for numerous rich discussions about trees, the Dharma, the Green Gulch landscape, and our role in serving through practice. I especially thank Davis Te Selle, whose art and collaboration on the book added a depth and dimension way beyond what I could have conceived of on my own.

Many places are mentioned throughout the book; these are the landscapes that have shaped my mind, body, and understanding of the natural world. I offer my thanks to Cave Gulch Creek and the western hills of Santa Cruz, Green Gulch Valley and the redwood canyon of Muir Woods, the high country of Yosemite, and the forests of Oregon and Washington. I am especially grateful to the California Academy of Sciences for the use of Pepperwood Nature Reserve. Their support for a month-long writing retreat made the completion of the book possible.

I thank my parents for the gift of life—my father, Eugene Kaza, for his passionate heart, and my mother, Nancy Snow, for her rigorous mind. Their love for life lives in me in the writing of this book.

I thank John Elder, Claire Peaslee, and David Abram for reading the manuscript in various stages and offering helpful comments.

Last I thank my editor, Joanne Wyckoff, for her strong and consistent dedication to the book from the early stages. Her attention to every word and detail was tenacious and fruitful.

May the joy and effort of this book serve the lives of trees, tree lovers, and tree landscapes and be part of the turning toward a vital and beautiful world.

Conversations with Trees

Introduction

THE FIRST TREE TO MAKE MY ACQUAINTANCE AS A child was the large-canopied apple tree in my backyard. Like many children I found tree climbing a favorite playtime activity. When our family moved to Oregon, I was fortunate to be surrounded by tall Douglas firs and other impressive giants of the Northwest conifer belt. My first teaching job in outdoor education placed me among coast live oaks, bays, and madrones as my daily companions.

For almost as long as I can remember, trees have been a significant presence in my life. I have studied them as a naturalist, slept out under their arching boughs, made tea from their needles, and preached about them from pulpits. I first began writing about trees as a theological experiment. I was engaged in personal spiritual development at Starr King School for the Ministry, and I wanted to understand how the gospels of love applied to nonhuman beings. I wrote the first piece during a Zen Buddhist meditation retreat, inspired by Martin Buber's thought on I-Thou relationships.

At the time I worked at the U.C. Berkeley Botanical Garden, where I was exposed to exotic tropical trees and unusual species from around the world. My interest in trees was spurred on by concern for the disappearing rain forests and loss of California oaks. Teaching children, adults, and undergraduates about the serious state of affairs for trees left me deeply disturbed about human-tree relationships. I traveled and taught in Costa Rica and Thailand and saw firsthand some of the damaged forests. I followed with dismay the politics and economics of old-growth forests in the Pacific Northwest, watching the checkerboard pattern of clear-cuts take over more and more mountaintops.

In each situation I saw the same conflicts over short-versus long-term gains, economic benefits versus ecological impacts of human activity. The greater the damage and loss, the more I questioned the justification for so much death. I could not make moral sense of the enormous waste and misuse of trees for consumer products. My scientific study of conservation biology fueled my ethical and spiritual concerns. Standard forest management practices did not seem to me to go to the heart of the problem. I began to realize that the social ethics of industrialized culture did not include any thoughtful basis for mutually respectful relations with trees. The predominant Western economic philosophy identifies trees primarily as objects for human use—whether for building, making paper, creating recreation sites, or producing toothpicks. This did not adequately address my own personal experience with trees and my growing sense of moral obligation to forests and woodlands.

To relieve the tension of this ethical dilemma, I joined others in local tree-planting efforts, discussing the biological and social aspects of restoration work. The work was part of spiritual practice at Green Gulch Zen Center in Marin County, California. Questions about relationships with trees took on a vitality of place and religious teaching that increased the complexity of the conversation. It was the power and momentum of this deliberation that spurred me to write about trees.

My practice of writing with trees is based in the Zen form of *shikantaza*—just sitting. I spent time in silence, close to trees, doing my best to be simply present with the tree as Other, aware of my thoughts, moods, and projections. I had no idea at first how this would work, but I persisted in the experiment. The writing became an excuse to listen for a call from the trees, in whatever form it took. I did not go to the trees with an agenda or story in mind, but chose rather to see what would unfold by being completely present in the specific place and moment.

This book is a chronicle of these explorations in human-tree conversations. It has been encouraged and supported by the sincere efforts of many people who are also searching for meaning and reciprocity in their relationships with the earth. The setting, for the most part, is in the coast ranges of central California, though I also sought out trees in Oregon, Washington, and the Sierra Nevada. The central coast trees are my backyard trees, loved through all the seasons as I walked the roads and trails of Santa Cruz, Mount Tamalpais, Skyline Ridge, Muir Beach, and Pepperwood Reserve. In this benign Mediterranean climate I could walk at night, in the rain, and throughout the dry summer, listening for conversations with trees. Tree habitats in California range widely from the open oak savannahs of the Central Valley to the deep conifer forests of the Sierra Nevada. The state is blessed with enormous botanical diversity, supporting over eighty species of trees. This book reflects the places I have been drawn to and the most common trees of central coastal California.

The context for the book is broader than my own personal experiences with trees. In this time of environmental crisis many people are responding with depth and moral concern to what they see happening. Upset by the impacts of a wasteful, materialistic Western society, many activists, teachers, and ordinary people are looking for more ethical and sustainable ways to live. For some the heart of this seeking is a journey of spirit.

A number of factors contribute to making this time very rich for spiritual inquiry in the realm of the environment. In this country the strong tradition and history of public lands protection, especially in the West, have fully matured by the end of the twentieth century. Millions of people camp, hike, travel, and seek contact with the natural world in national parks and forests. Beautiful trees and inspiring places are known by more people than ever before. Knowledge and concern for the health of the land have spawned hundreds of fiercely dedicated environmental groups, motivated by a sense of moral obligation to the extremely rich biological and cultural heritage of this continent.

At the same time an increasing percentage of Americans know almost nothing about the natural world. As more and more of the population becomes urbanized in physical location and cultural perspective, more people are psychologically and spiritually distanced from intimate interaction with the environment. I believe that television and nature movies contribute to this distancing by offering delusional substitutes for rich, sensory contact with the actual rhythms and textures of the natural world. Commercialization of trees, animals, and landscapes to manipulate consumers only adds to the distortion of environmental relationship.

Weaving through the American tendency to idealize nature, I see a strong thread of moral indignation and spiritual inquiry, based often on experiences of despair and grief over environmental loss. As more facts accumulate about toxic groundwater, nuclear waste, loss of ozone, and rates of forest cutting, one becomes most naturally overwhelmed by the extent of the damage. This can induce a state of denial, leaving one paralyzed to act. But for some the opposite happens. I have seen students and others experience profound moments of awakening to global interdependence. As their minds open, they see that the environment is everything. It is not just where we live; it is the very reason we are alive.

These moments of awakening are causing the American public to be increasingly dissatisfied with business and governmental decisions regarding severe environmental abuse. The environmental movement is, as much as anything, a struggle to reclaim the land and relationship with the land for the common people. Those with the greatest momentum to engage in this struggle are motivated by a powerful emotional and often spiritual identification with the environment. Political analysis falls short in explaining the force behind the evangelical dedication of people who have fully awakened to the interconnected reality of person, planet, and all beings. In the confusion around moral questions in our time, the natural world is a place of truth, generating ethical power by its very existence.

In this book I have taken trees as a place to investigate this naturally occurring truth. It is my sense that the root of ethical response springs from revelatory experience, the sort of encounter that penetrates to the core, illuminating one's perspective on everything. The power of this experience elicits awe, sometimes dread, sometimes unifying love. It is not something to be taken lightly. In meeting the trees I have asked for this direct encounter, wanting to be moved as deeply as possible by the trees themselves.

The pieces are investigative rather than prescriptive. In this regard I am strongly influenced by my Zen training. The experience of the truth of interrelationship or mutual causality arises independently for each person, exactly in the specific context of each unique mind and life. Thus the investigative process reveals the nature of reality in a way it can be most clearly seen by each person, with power to illuminate the individual's choice of actions. In contrast, prescriptive moral recipes for human-tree relationships tend to be oversimplified, cutting short the depth of transformation possible from thorough investigation. I have found that much more information about systemic relationships can be uncovered through an investigative approach, at both the individual and the social levels. The point is to rigorously and thoroughly examine the patterns that condition one's thoughts and actions regarding the environment, and in this case trees. Then one can make ethical choices to act appropriately, in harmonious synchrony with the dynamics of life and death.

My primary orientation in the book is not to tree as symbol, but to tree as Other, as one party in an I-Thou relationship. Trees have historically and mythologically represented many things—the Tree of Life, the axis of the earth, tribal ancestors, homes of spirits. But my effort here, awkward as it feels at times, is to try to speak directly with trees. I am more interested in one-on-one dynamic relationship with trees than in cultural constructions of trees or anthropomorphic projections. I am certain that I lapse into psychological projection in places, but my primary motivation is the desire for genuine contact at a core level. I have attempted to watch for the habits of language and mind that block the flow of communication between person and tree. These include stereotyping, objectifying, idealizing, and oversimplifying—all of which make a tree more or less than what it actually is. I have seen how fear, helplessness, and grief deflect movement, so I have made a practice of staying with the pain and intensity of these emotions to see what is revealed.

The relationship between person and tree, arising over and over again in many different contexts and with various individuals, is one subset of all human-nonhuman relationships. I am exploring it here as a way to inquire into the nature of these relationships. I want to know—what does it actually mean to be in a relationship with a tree? Acknowledgment of and participation in relationships with trees, coyotes, mountains, and rivers are central to the philosophy of deep ecology. In this writing I express one person's experience of the truth of this philosophy. By sharing this process with others I hope to encourage and support people in engaging in their own serious conversations with trees. In these meetings of tree and person I allow myself to see and also be seen by trees. I assume that sensing of the Other is two-way and active, though I cannot describe the biological basis of this for the tree. Over the course of writing these pieces I caught some glimpse of my meetings with trees as conversations in a co-created field of experience, generated as much by tree as by person.

I have freely experimented with different forms of conversation—from storytelling to dialogue, prose poetry to prayer. My desire has been to find ways to speak "with" or "to" trees, rather than "about" trees. Throughout the book I use the style of personal narrative to communicate directly the nature and details of my experience. I do not claim to speak for any others except myself; someone else meeting the tree in another circumstance would certainly find a different encounter. Yet at the same time I believe there is a strong core of truth telling in the process of participating in these conversations that may be recognizable and useful to others.

The framework of the book follows the central teaching of the *Mountains and Rivers Sutra*, a Buddhist text by the Zen master Eihei Dōgen of thirteenth-century Japan. The truth of this sutra, like most Zen works, is best realized through direct experience rather than cognitive explanation, and in this light I offer the book directly to the reader for their own insights. In brief overview, the sutra begins by acknowledging the simplicity of recognizing things as they are—mountains are mountains, rivers are rivers, and by extension, trees are trees. But in the course of studying mountains and rivers in depth, one sees them explode into all the phenomena that support their existence—clouds, stones, people walking, animals crawling, the earth shaking. Then mountains are not mountains, rivers are not rivers in the original sense. Proceeding with this investigation, one finally pierces through to the truth of the entirety of existence, including the mind of the perceiver. Then mountains are once again mountains and rivers are once again rivers, but now the depth of understanding penetrates, informs, and transforms the perceiver into a participant in the mountains' existence.

The five sections of the book unfold with this progression of experience. The first section expresses the simple desire to meet trees and make contact. I explore how it feels to be called by trees, to open to the Other in tree form. I follow various avenues of greeting—tracing the shape of the land, using my hands to touch trees, observing them with books and field guides, playing with children, and responding through the sensual. I take these trees as I find them, recognizing my own uncertainty and clumsiness in conversation.

In the second section I begin the process of looking past first impressions, taking the time to uncover more complete histories of individual trees. I find, for example, that trees tell stories of fire, agriculture, and commercial cultivation. I begin to look more deeply into my own needs for relationship with trees and how these influence my perceptions of community, change, and death. I encounter a greater sense of complexity in reviewing the role of trees as both victims and shapers of human activities. Some aspects of what I see are disturbing and unsettling. In trying to grasp the age and history of another being, I experience a certain vulnerability in raising difficult questions. With some tension I bump into the shortcomings of human capacities for deep and engaged relationships with trees.

By the middle of the book I have completely entered

the tangle of human-tree relationships. Trees are no longer simply trees; they carry painful stories of fear, killing, unconsciousness, and objectification. I sit with these faces of suffering, feeling the dilemmas of each situation. Answers, solutions, easy plans for fixing the damage are nowhere in sight. The trees explode into unending waves of despair, greed, and helplessness. I am caught in dialogues of time and place that reflect a long history of habits that distance and kill the Other.

Given the confusion and agony of this, the fourth section considers ways to respond that are heartfelt and genuine, that speak from the depth of what I have seen. I place my effort in cultivating a stable and attentive mind. Using time-proven religious practices, I aim for greater capacity in approaching the very demanding situation of trees today. The methods of renewal and awakening include pilgrimage, mindfulness practice, and spiritual inquiry—all influenced by the Buddhist tradition. From the natural world I look into the gifts of evolution and an animal body, and the quiet richness of the dark unknown.

Though none of the external circumstances have changed, one's internal capacity for action evolves through cultivating tools of consciousness. Rather than acting from a simplistic view of trees, I am now compelled to act from a context of mutual causality. In the fifth section I seek ways to restore spiritual as well as biological relationships with trees. In recognizing the dynamic working of mind and the many contributing factors to each tree story, my actions are tempered by the horrible, full, awful truth of the damaged planet. I look to the domesticated world for the obvious tasks of cleanup and healing. And I look to the wilderness for elemental teachings, for the power of evolutionary truths.

I have, in the end, no grand conclusions. The desire for conversation has only brought me closer to many difficult questions about reinhabiting place, living simply, and speaking from the truth of experience. As in most good conversations, there is the desire for more contact, more time together, and more depth. To cultivate this opportunity will require a level of love, effort, and spiritual integrity that I can only just begin to imagine.

PART ONE

Reaching Out

CHAPTER 1

Close to Water

SYCAMORES! BRIGHT WHITE TRUNKS AGAINST THE blue horizon! All this week I have been seeing you from a distance. Now I am lured closer for a conversation, for a chance to glimpse the world between your branches. Sweet billowy clouds dance above your arms in the piercing equinox light. The bright sun reflects off your tall trunks in the crisp, clean air. A two-day storm has pummeled the dry mountains and filled the creekbeds with the roar of spring. In the rocky Santa Monicas of southern California, sage and coyote brush glisten with rain. The young ridges rise up sharply from the coast, catching the morning light.

The sycamores lie in a narrow wilderness canyon two thousand feet above sea level. Downstream a hundred yards the creekbed opens into a grove of coast live oaks and bays—an unusually moist spot in this otherwise dry chaparral. I clamber along the footpath, following the movement of water. I can barely hear the trees against the loud rumble. The splashing and tumbling shouts a spring wake-up call to the watercourse down canyon. *Water! Water is here now! Drink while you can before the dry days of summer!*

In this beginning spring time the sycamore branches are almost bare. The wind off the ocean rustles through the unleafed limbs in company with the creek chorus. Despite this last round of rains it is another dry year in California. Five years of drought and I, too, am thirsty for a good drenching. A visit with sycamores is a chance to speak with water, for sycamores mark the trail of watercourses through dry and wet years.

Many creeks in this part of California are too ephemeral to support sycamores. The more common big-leaf maples and alders can make do with reduced flow or variable rainfall. But sycamores require a steady supply of groundwater. Sycamores are the great trees of riverbanks, of wet places, of fertile valleys, of high water tables. Their graceful, arching branches and sturdy trunks signify places of cool moisture, havens in the glaring heat of summer. Some large sycamores are still growing in spots where water once flowed but has since been diverted. Following sycamores is a sure way to know the path of water, the path of nourishment and connection.

Seldom does one see a perfectly straight sycamore trunk in the dry West. Mature trees are more inclined to lean against a convenient boulder or prop themselves up on the ground. The gnarled and oddly jointed branches show an uneven history of growth. Where limbs have broken, the trees compensate by sprouting clumps of canes. How much growth a tree can support in any given year depends on the amount of rainfall.

Of the five sycamores in this rare canyon oasis, four have achieved adulthood. One, not ten feet tall, has barely gained its sky voice. The thin, straight shoot is hidden by the thick branches of the giant next to it, twenty times its width. The large tree has several convoluted branches, twisting like snakes in the upper reaches of the tree; these mark the site of a former injury and the tree's effort to recover. It is as if the tree swallowed up the shape of the river and then let it loose to heal the wounded limb. And now the flow of water is the centerpiece of the tree, the place of testing and resilience. In healing itself the tree has made visible for others the healing power of water. With the glorious but short-lived abundance of seasonal water, today's sycamores seem filled to the brim, strong and present, already surging with the life force of spring.

Sometimes you have to look at a tree from the ground. On your back. Nothing between you and the sky except the arms of a tree. Just lying on the earth, looking up. This is a fine posture for listening to the wider conversation of native voices. I gaze up at the jigsaw pattern on the trunk where pieces of bark have fallen off. It looks as if the skin can't grow fast enough to accommodate the widening trunk. One of the trees is almost completely peeled to a smooth, white skin. Another is pockmarked by splotches of dark gray, light gray, and tan. Shards of bark lie jumbled at the foot of the trees in a random pattern. One by one the pieces have dropped here, creating by chance a statement of beauty few will notice.

Usually I am drawn to sycamores for their porous canopies of maple-shaped leaves. The broad leaves arrange themselves to expose the greatest surface area to the sun, thereby increasing both photosynthesis and moisture-preserving shade. Over and over again I have been lured to the cool respite of sycamores and the mesmerizing play of their graceful shadows on the ground.

But there are no big leaves yet, nor any flower strings of fuzzy golden ornaments. I see right through the skeletal branches to the open sky. Only an assortment of spherical seed pods and a few shriveled leaves decorate the trees' empty arms. The newborn leaves must have been just unfurling out of their bud cases when a cold front caught them by surprise. The tiny leaves hang limp and lifeless, soon to be discarded when the trees try again under more favorable conditions.

Against the backdrop of unstoppable spring ebullience I am struck by the vulnerability of new leaves. It is a strange paradox that the force of spring begins in such tender and fragile life-forms. Water is apparently not the only factor in sycamore success. Timing is also critical. Deciduous trees must make educated guesses as to when the winter cold fronts have passed and it is safe to send out new growth.

Each leaf becomes a calculated risk of time and energy, a capital investment in photosynthetic potential that will serve the tree through summer. The sycamores, like most trees, create these first leaves from the last of their winter food stores, knowing that conditions for growth improve with the lengthening days. Thus leaf production is an act of faith based on right timing.

In the face of the naked vulnerability of shriveled leaves I am surprised to feel the sycamore's strength. The trunk is sturdy, easily supporting my weight. Yet I do not know how this tree's story will actually turn out. It is not a foregone conclusion that this sycamore will produce healthy leaves every year. In dropping close to water, I begin to see the unpredictability of the ebb and flow of life force. Extremes of flood and drought may stress the tree's capacity for growth. A sycamore must be prepared for a range of conditions far more challenging than this kind spring day.

Yet the resilience of trees is something I have always counted on. It supports a certain confidence that life will go on, noticed or unnoticed. I rely on the assumption that sycamores will find water wherever they grow and produce extravagant shade against the summer heat. But truthfully, I do not know how much water sycamores need to stay healthy or how this varies from tree to tree, depending on size, location, soil, and slope face. I do not know if these trees are close to their drought limits or struggling with accumulated stress over a number of years. I feel the pangs of ignorance and uncertainty.

In the last century ninety percent of the beautiful sycamores up and down the California riparian corridors have been lost to land clearing, river channeling, and housing development. I have come to think of the remaining sycamores as reminders of another, more gracious era in human history, when people sat out on their porches enjoying the summer shade. I want to believe there was a time when people were not so quick to destroy the trees that nourished their lives. But all that has changed, and sycamores today are subject to cumulative hardships not only of reduced water supply but also of agricultural pesticides and air pollution. I fear the subtle stresses will cut short my conversation with sycamores, reducing the possibility of long-term relationship.

As water trees, sycamores reach deep into the ground, looking for water even where no stream is visible. It is this reaching into the unknown that beckons me in the uncertainty of these trees' lives. I want to come close to water as it percolates through the soft soil, licking at the root tips of thirsty cells. I want to follow these water markers across the dry landscape, finding a route of shelter and nourishment. I want to know something about long-term relationships with trees that last over centuries. I want these sycamores to tell me stories from the living stream of their history over millions of years.

In a time when fragmentation dominates the landscape, I need to meet these trees in sycamore context, investigating the wholeness of their lives. Can I be bold enough to step aside from what I know in order to be available to what I don't know? Can I follow the meandering snake routes of water to ask for the help of trees? In the paralyzing rush of all that moves fast, I want to meet these sycamores in the slow time of water.

CHAPTER 2

Called by Alders

INSIDE THE MEDITATION HALL I SIT WITH MY HANDS in my lap, breathing into the hollow of my palms. Left hand lying gently inside the right, thumb tips just barely touching, center point of concentration in the posture of Zen meditation. Back straight, mind deep in the body, the hands open and connected to the motion of breathing. I think of the image suggested by the Vietnamese Zen teacher Thich Nhat Hanh—a baby Buddha nestled in my hands, a radiant being in the company of my hands and breath. Let the hands be taught, let the hands absorb the Way of the Buddha and teach the mind.

Outside, the winter creek dances through the rocks, refreshing this small hollow below Skyline Ridge. Evergreen bay laurels cover the quiet spot with shade and whispering. In between periods of meditation I walk slowly around the deck, drawing in the sweetness of early spring. By mid-morning the land is speaking more powerfully than the meditation hall. *Come, come, walk with me, smell the fresh soil, catch the bright sky, come to the trees, come.*

The fire road above the retreat center leads down to the pond. From there it is a short walk to the ridgetop. The ridge stretches for fifty miles from San Francisco south to Saratoga Gap, with wide views of the rolling hills on both sides. I head out in my black robes, leaving behind the strict schedule, choosing instead the irresistible pull of the land. Deep, long breaths, each foot touching the earth in slow, rhythmic strides, connecting small body to large body as I walk.

I have barely entered the forest when I see the alders across the pond. The invitation to stop fills my attention; I am riveted by their graceful beauty and still reflection. Alders, I see you from this open spot by the tall Douglas fir. You are very quiet this morning. The light touch of a breeze tickles the pond's surface, rippling the reflection of your white trunks and open branches. Early April is such a tender time for you—all your fresh green leaves just out, catching their first taste of sunlight. I wonder how it feels for you to stretch out your green hands to the light after months of empty branches.

From this cordial perspective I am taken by your remarkably clear and detailed reflection. The image forms perfectly in the first quiet hours of the day when the surface tension is light and dynamic. By late afternoon the surface will be thicker and less responsive to subtle breezes. The pond will fall into shadow, and insects will stir across the face of the water. Right now—moment of mindfulness, right now—moment of impermanence.

Alders, can you feel the beauty of your reflection? Of course not. And yet, this great beauty lies right at your feet. Are you not connected to it in some way? Does not each shining being contribute to the elegance of this entire event? I must come closer. I walk around the lower berm past the cattails.

Now in your presence, I see how straight and elegant you stand, especially next to the gangly willows. Your gray-white dappled trunks are tall and smooth, as if you shot up fast before any injurious forces could do you harm. Your trunks are barely scarred except for the knots from past years' branches. I extend a shy hand, meeting your body, firm and cool against my skin.

I wonder if a tree knows when someone's hand is on its body. Does it feel a little warm, like an exchange of electricity? This act of reaching out is a small gesture, but it is filled with great intention. I am simply trying to say hello across the barriers of form and language. I believe the hands communicate this intention most honestly. The voice and mind are not direct enough. Or perhaps they are too complex for the first step of making contact. Besides, the tree and I have such different minds and voices. I don't know the language of these alders at all. I can only guess at the shape of a tree's mind and what it knows about life on the edge of a pond. How does this water taste to an alder? How does the morning sun feel on its new leaves? How does the wind feel moving through its branches?

I know a little more about the pond from swimming its warm waters in summer. Created originally for water storage, it is only a hundred yards long, thirty yards wide—a filling in of a natural draw through the hills. I can swim from one end to the other in five minutes or lazily cruise the perimeter in twenty, checking the mud spots for frogs and turtles. The banks have filled in with willows and cattails, providing hiding places for mallards and marsh wrens. Many times I have heard the sharp rattle of a kingfisher piercing the air overhead.

The prospect of a swim is tempting, but I resist. Today only my hands will go for a dip. Down they go in the still water, shattering the reflection. Down they go, reaching for alder roots, bright red alder roots. Like slender snakes swimming in a tangled mat or my long hair floating behind me when I swim, the roots drift in the water, soaking up the life-giving liquid. Slippery, smooth, wet—how soothing it must be to feel the water move between your toes! The little rootlets sway back and forth in the gentle stirrings of the pond, drinking the sweet water through their root hairs.

Hands in the water, hands on the firm gray trunk, hands exploring the shapes of these trees. I lean out to feel the season's new leaves—smooth on top, fuzzy below, rough and serrated at the edges. The flowering catkins, now past their peak, still dangle in the shiny canopy. A month ago, the alders bloomed near my house on the coast. The trees were decorated with hundreds of catkins dancing in the wind. The finger-length male catkins were explosive to the touch; if I knocked them lightly, a big poof of yellow pollen would go flying.

Pollen is key to an alder's reproductive success. Alder flowers, like most flowers, are vulnerable to weather and the potential for missed timing. Deceptively small, less than an eighth of an inch across, the compact rows of tiny flowers can easily become bedraggled in a storm. If it rains hard for several days or weeks after the male pollen matures, all that pollen washes away in the stormwater runoff, never making it to the female flowers. A season of seeds is cut short.

With each week of the new spring season, the alders move further into full greening. I wonder how these waves of productivity feel after the quiet time of winter. Does the new activity make you tingle? Do you experience the returning of the long days of sun as pleasant, as dynamic?

It is difficult to imagine the scale and complexity of activity that goes on inside a tree. I sense in my hands some charge, some energetic force at work here. Joining palms to trunk, I form a circle of energy with this tree. Listening through my hands, I meet this tree from my own experience of sunlight and stillness. Bound to a single place, perhaps the tree delights in the simple rhythm of the changing day. Maybe it is similar to sitting in the meditation hall, watching the day go by when there is nothing else to do.

First the day begins, then it unfolds, and soon it closes down, and the night sky appears.

Day after day the rhythm repeats, shaping the tree's life through the seasons. The tree itself is a manifestation of rhythm, of the way the light works. Supple trunk, dainty flowers, fresh green foliage—all are testimony to this pattern of change. This is what my hands recognize—this movement of sun, earth, and water. At some fundamental level this tree and I are made of the same rhythms. We share a common understanding, available in the meeting place of touch. Reaching out, I find a simple way to begin a conversation. Coming close, I offer my hands in greeting.

CHAPTER 3

Red Fir Encounter

ITCHY FEET THIS MORNING; TODAY I LEAVE FOR THE High Sierra. *Mountains of majesty, mountains of splendor!* I am caught up in anticipation, eager to see the great trees of the high country. Like some territorial animal I yearn to roam the distant range and leave my mark on the landscape. I need to leave the world of complication and enter a world of simplicity. I am more than ready for the change.

Cookbox, stove, tent, and sleeping bag. I consider each item carefully—what do I actually need? Water bottle, matches, wool hat—the choosing becomes a ritual of stripping down, quieting the mind from endless distraction. Like most mountain expeditions, this is a journey for the soul, a seasonal rejuvenation of spirit. I long to taste the sweet mountain air, I long to drink the intoxicating water of high alpine creeks.

I know from experience that the mountains and their

trees carry far more spiritual presence than the concrete jungle in the city. Like thousands of other campers and hikers I find that chances for profound encounter increase in the unencumbered spaciousness of the high mountains. Here in the legendary Sierra, home of giant sequoia, sugar pine, and incense cedar, I can wander with fresh eyes, ready to receive the teachings of trees.

The last thing to pack is the library box. My collection of Sierra field guides has swollen in recent months with the addition of a large number of tree books. I want to know about the rich diversity of trees that meet here. The field guides will be my tutors in understanding how trees from six climatic regions have come to this landscape. On the western flanks I'll look for gray pines, sycamores, oaks, and bays from the mild coast ranges. Dwarfing them will be the tall conifers of the California mountains—giant sequoias, Jeffrey and sugar pines, red firs, and incense cedars. From the cool, wet North Pacific coast I'll recognize Douglas firs, mountain hemlocks, and Pacific dogwoods. From the northern Canadian boreal forest come lodgepole pine and quaking aspen, from the eastern Rockies ponderosa and whitebark pines, white firs, and water birch. And on the eastern slope live the dry trees of the Southwest and Great Basin—foxtail and limber pines, piñon, and juniper. I want to walk among these trees and look for evidence of their journeys from distant ranges. I suspect that what I know from books is merely a glimpse of what I might know through my feet and eyes.

I fit the last sleeping bag into the car, check for maps and sunglasses, and finally the ritual is complete. As a twentieth-century religious ceremony it is notably imperfect, but there is power and motivation here all the same. I offer a small bow, pray for safe travel, and start the engine.

The five-hour transect across the state dips up and over two coast ranges and across the flat, open Central Valley. Past the almond orchards and fruit stands, across the Sacramento River, I wind toward the first low humps of the foothills. In the heat of the day the dry grasses are like waving matchsticks in the oak savanna and chaparral. Among the oaks stands the first of the Sierra trees, a gray or Sabine pine. In the bright sun of midsummer the long-needled, airy tree shimmers against the cloudless sky.

At the top of Old Priest Grade the thick cover of chamise and monkey flower gives way to open stands of ponderosa pine. According to the field guide, this tree is found throughout the western states, preferring level or rolling dry land. The ponderosa pines keep a certain distance from each other, creating a gracious beauty characteristic of the dry forest. The inviting spaciousness is the result of competition for water; each tree's root system competes for snowmelt percolating through the pine-needle duff. I find myself moved more by the aesthetics than the science of these agreeable trees. Their cinnamon-orange boles rise splendidly in the parklike setting, capturing my eye and thus my heart. These signatory trees of the mid-elevation Sierra mark the entrance to the mountains. I inhale deeply the dry resinous incense and begin the purification of body and soul.

The road climbs up through two thousand feet of pines, black oaks, and Pacific dogwoods, their white platter flowers dancing in the shadows of the forest. At Tioga Pass I take the left turn to the high country, leaving behind the popular waterfalls and glacial domes of Yosemite Valley. After climbing to seven thousand feet I pull off into unfamiliar territory down the old Yosemite Pass Road in search of a campsite. The four-mile road is slow going, with eroded potholes, rutted turnouts, and narrow one-lane passages. Under the intensive pressures of San Francisco Bay Area weekend warriors, most of Yosemite's campgrounds are reserved throughout the summer. But this out-of-the-way spot is relatively empty. With great delight I find a scenic, isolated site next to the river and make camp.

Here—am I here yet? I need to walk the Sierra back into my feet to really arrive. With a daypack full of field guides and binoculars, I set off down the road for the meadow of shooting stars I passed on the way in. I feel the urge to range over the granite landscape; I find myself shouting to the noble trees. *Hello, noble trees! You are magnificent!*

The meadow is flamboyant with wildflowers, a dazzling palette of pinks, blues, purples, and yellows against a sea of lush green grass. Tall pink and white shooting stars spill across the meadow like snow drifts. Blue camas lily, yellow groundsel, pale Sierra rein orchid glisten in the midsummer light. One part of my mind wants to identify each of these delightful beings scientifically by name; another wants only to gaze at their exquisite beauty. Two approaches to the marvelous complexity before me, two ways of rejoicing in form.

Boggy and wet, the meadow for all its beauty is too sloppy to explore. I retreat across the road for a landscape painter's view. Leaning against a large tree trunk, I realize these are red firs, the statuesque, symmetrical trees of the snow flats I read about. *Abies magnifica*—magnificent firs. Deep chocolate-red coats of furrowed bark, stately branches in the somber darkness of the forest: even the fallen trees carry a grand presence.

I lie back on the forest floor, in no hurry to leave the company of these red firs. The canopy is a visual feast of filigree lacework. Side stems off each branch grow evenly spaced, almost to perfection. Each cluster of needles curls upright, giving the whole tree a vital and resilient appearance. The simple elegance of this pattern is repeated over and over, branching and lifting, branching and lifting. The pure evolutionary lineage of tree intelligence overrides all my ideas of beauty. In one moment of unity everything lines up, and I meet the red firs with a thundershock of connection. For a moment I am almost paralyzed.

Who *are* these trees? The field guides don't say much about this kind of identification—identification with Other, meeting the tree as fully itself. Names and range maps don't capture this depth of feeling. I am temporarily lost in the timelessness of red firs. The trees have unsettled me.

In an effort to get my bearings I try to compose a story of their lives. At my feet hundreds of seedlings pop out from the litter of fallen branches. Most of these inch-high trees will not make it past this tender stage; the first winter snow may be too heavy for the new growth. Those that survive will grow to look like the three- to eight-foot saplings across the road. Though they have survived a few snowy winters, they are still youngsters, not yet adolescents. By the time they reach thirty feet, the young adults will sport their first cones. I try to imagine these groves yards deep in snow, the firs wrapped gently in a soft blanket of protection from cold and wind. The most successful trees will mature at two hundred and fifty years, like the sturdy giant I lean against. When the trees fall, they will decay quickly on the moist forest floor.

Between the field guides and the growth stages before me, I am trying to piece together a red fir's life history. It is a rudimentary exercise, this construction of pattern from fragments of information. As an infrequent visitor to this forest I cannot know the story of a single tree from birth to death. My understanding is limited to what I can observe and how I make sense of it. The moment of connection presents a framework, a place to link what I know with my experience of direct encounter. The storytelling is one way to ground the shock of deep meeting in the familiarity of relationship. I take the pieces of the story and invite myself in to the rich domain of red fir history and relationship.

I didn't know this would be the point of encounter; I hadn't truly met the red firs before. In the process of stripping down and simplifying, I made room for a wider reality. The fullness of this feeling includes an edge of obligation. Meeting deeply, even for a moment, confers the responsibility of knowing to the knower. With these firs I am given the task of searching for insight through science and beauty. Lifetime work, I realize. In a single moment the red firs caught my mind and changed the shape of my internal landscape. Now I am left looking for details to flesh out the larger story.

CHAPTER 4

Friends of the Family

DEEP STILLNESS OF REDWOOD GROVES. THE BREEZE barely penetrates this quiet spot where I have come for companionship and a story. A few jays and warblers chatter in the distance on the edge of the meadow, a towhee scratches in the brush. In the soft afternoon light there is little trace of the silent drifting fog that nurtures these trees through summer. It rolls in off the oceans and up the hills of the marine terrace in a great sweeping rhythm. In at morning, out at midday, back by evening. Sometimes the fog lies heavy over the forest all day long, spreading inland to cool the Central Valley. The ebb and flow is at its pause point for the day. The air is very still.

Today I heard the calling of the redwoods again. It often comes when I am with children I love very much. A tender place in my heart wakes up to the voice of redwoods and I feel called to a conversation. I cannot explain this pull to the tall ones, but it is familiar and compelling. The redwoods that live in my memory are saying, *Come, come be with us today, touch us, sit with us, listen to us, stay for a while, let's enjoy each other.* In the sleepy afternoon warmth of mid-spring, I bring the sweetness of children here to these trees.

Children, please meet my friends, the redwoods. Redwoods, please meet my dear friends, the children. May we stop and visit with you for a while? Would you be so kind as to welcome us? We would like to get to know you. You live everywhere around us and yet we have seldom stopped for a visit. You are our forest, our neighbors, the shape of our landscape, but we haven't properly met.

On a shaded trail leading into the forest, a deer has joined the party. From a low branch a hermit thrush trills his silvery song. Bright triple petals of trillium poke out from the dark green leaves and shaded forest floor. A banana slug oozes along at a snail's pace, leaving a track of mucous slime. Sword fern and huckleberry complete the convocation. In a rapid burst of staccato notes, Wilson's warbler announces the gathering.

Friends of the forest, please join us; redwoods, please allow us to join you. We are delighted to be with you for this short moment in your lives. Today I have invited two children to visit with you. These young girls have met you before on our picnic journeys to backyard trees and in our magic stories told snuggled under the covers together. They

have been on many adventures with me and understand the power and delight of stories. They are still young enough not to have forgotten the truth of friendship and conviviality. I am hoping they can find a story with you, that you will share something of your life with them.

Look, children, the trees are growing in a circle, young ones side by side with the old ones. The circle marks the spot where a giant redwood once grew. When it died long ago, it left a ring of new trees that sprouted around the base. Now these trees have grown tall, and another generation of youngsters is sprouting. A baby tree living off a big tree can survive much better than one grown from a seed. Look at these tiny seeds scattered on the ground. They are less than an eighth of an inch long, 100 to 150 seeds in one small cone. Seeds have to find their own water, while a stump sprout can draw water from the parent's roots.

Do you see the green female cones hanging on the trees? At the tips of the low branches are clusters of much smaller male cones. The pollen from the male cone has to land on the female in order to make the seeds. The tree is always making seeds, for at any time a fire might sweep through, clearing the ground for seeds to sprout. How many seeds do you suppose have fallen on this little patch of ground?

But that's enough biology for now. The children need a chance to talk directly with these trees. I lean against the flared base of an old one, looking up at its formidable trunk. How *does* one address such a long being? Should we speak to the crown or the trunk, or to the massive roots? Where is the energy center of the tree? This business of introduction is awkward and confusing. When a person is so enormous, it is hard to have a proper introduction, much less conversation. Perhaps a smaller tree is an easier place to start, especially for children.

Hello, young tree. Your branches dance lightly in the breeze; your needles pulse with the surge of spring growth. No pollen or seed cones yet, you are still a child—only three feet tall. Little tree, you seem about the same age or perhaps a year or two younger than these children. Can you tell us how it is for you to be a child of the redwoods? Do you play? Do you like it here? Do the older trees take care of you? The children want to know about your life in terms they can understand. They are curious and friendly, willing to cross the borders of form to meet you as an equal.

As in all meetings with strangers, it is a challenge to find a common language, but it is not impossible. For children it is a natural request. Why not have tree friends? For a young one all beings are of equal interest and value until proven otherwise. Each greeting is a point of contact, a way to enter the conversation. Each introduction to salamander and shrew is a step toward seeing the pattern of lives and movement that define the forest. To say hello to trillium and redwood sorrel is to meet the friends of the tall trees, to come a little closer to the richness of the redwood community.

Hello, huckleberry. Hello, sword fern. Hello, varied thrush with your black necklace and red breast. Hello, brown creeper climbing up the tree. Hello, hairy woodpecker, *rat-a-tat-tat*-ing on the oak. Such a lovely gathering of children and friends on a sweet, sunny afternoon. What else to do but have a tea party? Let's play by this stump and make a house in this hollow. We can put the dishes over there and the tea here, and don't forget the acorn cookies. I'll sit on this log and you sit on that stump. Now, won't you have some tea? Thank you, it is delicious. Please have some redwood sorrel. Oh, it's sour! Don't we have a lovely house here? Yes, I think Mr. Jay likes it very much. Look how tall our house is! And here, where the trees make a tight circle, we can run in and out and catch each other and play tag. You're it. No, you're it. The tree is it! Don't let the tree touch you. Watch out! Oh, gotcha. Snagged by a branch. Oh tree, you got me.

Oh, I'm tired. Let's lie down here on the soft ground and get comfortable. We can move the big branches around and hollow out a good spot for our bellies and find little branch pillows for our heads. Ah. This is a nice spot in the center of the circle. Here we are together, all of us children in the cradle of your ancient presence.

Redwood elders, human children, you seem to be friends already. The children lie entangled in your fallen branches, at rest in the bones of your history. People and trees have met many times like this before. It is a natural friendship. Children are easy with you. Look, they have fallen sound asleep! Perhaps they remember without thinking, without doubting, the long-standing relationship between trees and people. Our good times together go way back—trees are old friends of the family. The children seem to know this. Soft faces in the streaks of sunlight, soft backs curled into the forest floor. Sing to these children as they sleep, dear redwoods, sing to them of your ancient ways and your long journey through time.

CHAPTER 5

Maple Ecstasy

IT RAINED LAST NIGHT—A SOFT, SWEET APRIL RAIN, caressing the earth with life-giving moisture. I was dreaming about bigleaf maples and their jubilant flowers of early spring. I awoke this morning filled with silky memories of maples—the tender new leaves pushing out of their sheaths, the pendulous lures of yellow-white flowers floating against the open sky. The flowers appear like jewels on a beautiful princess, calling suitors to come dance in the tangle of stamens and cover themselves with the delight of pollen. Sensual memories of maples along streambeds, maples in canyons, maples in the dancing light of midday gleam in my mind.

Soaking up this sweetness, I savor the delicious treat of this dream. As these trees touch my mind, spring fever surges in my flesh and bones. *Maples!* I am gripped by an intense yearning to see you, to greet you, to touch you, to be touched by you. Spring lover, you have come calling in my dreams, catching me in a moment of intimacy. I must go find you.

With the speed of a hummingbird, I dart into my clothes and buzz onto my bicycle. No quiet time this morning, the maples are calling. I race down the hill, shouting hallelujahs to the sky, the creek, the season of ecstasy. It is sheer madness to careen down this hill at breakneck speed, but the road is empty, it's early Sunday morning. I have the whole riotous place to myself.

The creek bank is splendid with alders—a lacy thicket of water people gathered in the cool canyon. Shiny new leaves cover the dangling catkins, cloaking the white-barked trees in excellent spring finery. I peer among the alders for the star-shaped leaves of maples, but they elude me. An occasional buckeye catches my eye, but in contrast to maples, the buckeye's canopy is round and compact, its trunk upright and predictable. I am looking for a shape more delicate, asymmetrical, even gangly. I am looking for the way maples lean for the light.

I scan the landscape for visual cues—the lilting angle of maples, the green-red color of new leaves, the quality of light flickering through the canopy. This anticipation is a sweet delight. The certainty that the inner and outer visions will soon converge is the joy of seeking the lover. Each gaze into the landscape, each query—Is that you? Are you the maple I seek?—is part of the process that builds the bond. Each glimpse is a statement of desire, a call for response from the lover-to-be.

Spring madness pours out of the oaks and bays in liquid streams of birdsong. Each melody I recognize adds to my intoxication. Orange-crowned warblers tickle my ears with long, descending trills; song sparrows chirp cavalierly, announcing their desire to mate. Two ravens call raucously overhead. As I cruise along the empty road, an Anna's hummingbird darts up to investigate my bright jacket, flashing his red throat. Hello, bright-spirited hummer, are you my lover today?

Around the bend I spot the season's first elderberry blossoms. Distracted from my search for maples, I stop to smell the balls of creamy-white flowers, blooming weeks before the buckeyes. Elderberries and maples produce some of the earliest tree flowers; bays, madrones, and oaks will flower later.

Many people never think of trees as having flowers because the flowering period is easily overlooked and often very short. Late-blooming flowers would be obscured by a thick canopy of leaves, so many trees produce their blossoms before the leaves, when they are more visible to insect pollinators. Maple flowering can be over in two or three weeks. By the time the leaves have fully matured and hardened off, the flowers will have been pollinated long ago and the first seeds already formed.

This narrow window of flowering time only adds to my anticipation. I am eager to catch the dainty buds and jewel green leaves at their peak. It is this exquisite event that will renew my love affair with maples. I don't want to miss it.

I spot two maples some distance across the creek. Ah, the first glimpse of my beloved! The delicate blossoms grace the trees with a pale yellow glow. I can almost taste the particular sweetness of sitting under a maple's arching branches, the mottled sunlight warming my body like a lover's hands. I pedal past these two trees and enter the deep shade of the redwood forest. The tall giants stand still and steady along the tumbling creek, framing the groves of silence. Now I must slow down and leave the bicycle behind. It is time to walk quietly toward the lover, calming the rush of anticipation before the moment of greeting. Too much eagerness and excitement will jumble the space of contact. Let it go, let the breathing even out, let the feet touch the ground. Let the body and mind prepare for meeting the Other.

My heart is pounding from the pace of bicycling. It seems foolish to be so taken by a tree. I would not confess to most people my mad crush on maple flowers. They might find my taste for maple ecstasy a little odd, perhaps even unsociable. If I love trees so much, they might reason, do I love people less? No, I respond, the two serve each other well. One expression of love deepens the other by cultivating the capacity to meet another being. This is no easy task. Walking along the creek, I am slowed by the prospect of realizing this intimacy.

Finally I stop. Elegant moss-covered maple—I see you. A great surge of joy ripples through my body. At last the point of contact, the pure physical delight of the thing itself—richer and more sensual than any dream or memory. The big-leaf maple arcs overhead in a gracious loop of circular energy sustained by the ground. The trunk's life force shoots upward, nourishing myriad new buds and flowers. Along the sloping branch, ferns drip with spring moisture from the night's gentle storm. I feel such tenderness in coming close to these exquisite flowers. Usually the delicate petals are out of reach. But after last night's rain a few strands lie fallen on the ground. I pick them up and hold them gently in my hands.

Sweet flowers, I cannot say why I love you so much. Each is plain and simple—the male flowers just ten small stamens in a modest circle less than half an inch across, the female flowers a single ovary. No brilliantly colored petals, no exotic eye-catching pistils, no outlandish masses of pollen—just very ordinary, less-than-astonishing flowers. And yet the entire cluster of blossoms is quite pleasing to the eye. Some of the female flowers have already been pollinated and the brick-red seeds are beginning to grow. I can hardly see these specks of red: they are not much bigger than a grain of rice. In a matter of days the whole tree will be covered with tiny seeds. These will store the first food from the new leaves and grow to full winged maturity under the cover of a luxuriant canopy.

Perhaps the fragility of your delicate life is what attracts me. The desire to be in your presence is the desire to know the beginning of things—the first true taste of love, the sureness of pulsing energy, the power of intimacy before the canopy obscures the reproductive urge. In this first delight of growing, there is an unmistakable urge for contact—pollinator to pollen, pollen to female pistil. In this contact lies all the potential for new life and genetic variation. The possibilities are tantalizing, the raw power of creativity undeniable. The density of energy in this bundle of flowers is magnetic, warming my hand with a small charge from the tree's electrical field, calling forth in me the same desire for creativity.

As the sun streaks through the canopy, a winter wren bursts into song, its warbles and trills cascading in the deep glade. Sing, little bird, sing! You are my voice in the company of maples, saying what I cannot say in words, singing the rapture of springtime intimacy. Like the maple flowers, you are small and modest, the tiniest of the redwood forest birds with barely a stub of a tail. But your unending song carries the spell of spring fever into the heart of the trees. At my feet a banana slug creeps quietly into the patch of fallen flowers. Two tempos of greeting—the ebullient wren, the plodding slug. I am somewhere in between, torn by two impulses: the adrenaline charge of spring fever and the liquid desire to melt into the presence of the Other.

The desire to merge slowly and imperceptibly is the deeper impulse; adrenaline merely initiates the motion. I have been propelled into the company of maples, irresistibly pulled by the desire to meet. Now, in the actual presence of the tree, I soften with the tenderness of the dance between two beings. I recognize this consummation of yearning, this fulfillment of desire. Sweet, graceful flowers, tender new leaves, let me be touched by your presence, let me be moved by your beauty.

PART TWO

Tracing the Stories

CHAPTER 6

Reference Point

SURROUNDED BY A SEA OF FOG, THE PEPPERWOOD bay tree stands like a sentinel on the edge of the ridge. Even in this dense ocean mist, I can see it looming from a distance. It is one of the most conspicuous trees on the three thousand acre Pepperwood Nature Reserve. The nearby coast live oaks pale in comparison, their foliage less dense and less massive. In the middle of the dry summer season the fog blanket has soaked the grass and given the hard-baked soil a taste of moisture. Just walking one hundred feet through this grass, my shoes and pants are soggy. I've entered the fog; the fog has entered me.

Under the tree a small rainstorm drips loudly from the branches, forming a patch of very wet ground. The large bay creates its own mini-weather system, changing the physical conditions in the immediate neighborhood to its benefit.

This is no small rain front under the tree: the full canopy stretches at least one hundred twenty feet across. Six large trunks grow out of a single bole, which is itself twenty-six feet in circumference, or the equivalent of ten eight-year-olds touching hands in a circle. The tree is massive.

Bay laurels, or pepperwood trees, often grow in clumps or clusters, usually as part of larger groves or woodlands of mixed bays and oaks. This tree, in contrast, stands completely alone. Surrounded only by grasses in an open savanna, it is an oddity on the landscape. The pepperwood bay is an island, a tree island of significant stature, a reference point for birds, people, and animals passing through. Cows are drawn to the tree for its cool shade; crows and red-tailed hawks perch here for convenient views across the valley. For me the tree serves as a useful landmark, for I have seen it from several vantage points. It helps me get a fix on the landscape. The locals say you can spot it with the naked eye from as far away as Mount Saint Helena, fifteen miles distant as the crow flies.

With the assistance of this tree I am building a map of the region in my mind. This is how it works: one starts with key reference points that establish the shape of the land. Peaks, ridgelines, craggy rocks, or large trees like the big bay all work well. One notes these landmarks from several different perspectives. The eye teaches the mind: this is where things are. Then, over time, one can walk from one reference point to another, filling in the details between landmarks. Each link marries the land to the mind, so one comes to know one's way around in a place. It takes a long time and a lot of walking to paint the full picture. Each eye-foot connection is a piece of knowledge, a cultivated intelligence based directly in the landform. This is the intelligence of the native dweller for whom sense of place is a reference point for all actions.

But now that I have stopped to meet this reference point up close, questions arise: Why does the tree stand out so dramatically on the landscape? Why is it so alone? There are no other bay trees anywhere near it on the ridge or over the dip into the next hollow. Yesterday I sat under the tree with the reserve's resident biologist, and we conjured up its story.

The pepperwood bay is a recent landmark, formed as an artifact of history. Core samples of bay trees have shown that trees twenty to twenty-five inches in diameter are 160 to 210 years old. This tree is at least four times that wide, so it may be five hundred to eight hundred years old, or even one thousand! Tree age does not necessarily correlate with size, but clearly this tree has lived awhile. It has, no doubt, seen many fires and survived many winter storms. We conjectured that once this area was covered with a more extensive forest of maples, bays, oaks, and possibly conifers. This forest, like all California forests, was subject to small and large cleansing fires. Every ten or twenty years a ground fire probably came through, clearing the twigs and litter below the tree. These may have been set by native Wappo and Pomo people using brushfires to drive out small game for hunting. Then every one hundred or two hundred years the forest might have been charred by a major conflagration, toppling old trees and scarring young ones, changing the shape and character of the entire area.

We guessed that the original tree survived the small fires relatively unscathed and grew to develop a substantial trunk, the base of which is still visible. Then in a major fire it lost its crown down to about six feet above the ground. After the fire the thick base sprouted, sending new shoots of growth up around the edges. These in turn, over the next several hundred years, grew into full-sized individual trees, creating a small grove out of the stump. This tree, for all we know, may have actually lost its crown and resprouted more than once.

By the time the first wave of white settlers came to the area in the 1800s, the tree was so large at the base that it probably stood as the grandfather tree on the hill, wider and more prominent than the trees that came after the fires. For the colonizers the land was a more valuable commodity than the forest. While the native Indians depended on the trees for acorns and bay nuts as primary sources of protein, the settlers were interested only in the soil and what it could grow.

I try to imagine this conflict of use and perspective, knowing full well which point of view prevailed. County records document the step-by-step fragmentation of the original forest. Like gold in the Sierra foothills, grapes were the currency of consequence in the coast ranges. Almost the entire forest around the bay tree was laboriously cleared to make way for vineyards and the shiny promise of a cash crop. Row after row of grape stakes and fence posts were pounded into the ground, cut from local redwoods. A winery was built, and grape products became the major source of income in a booming local business.

Meanwhile the grandfather bay tree stood in place, too big to tackle and not worth the effort. Other stumps were rooted out or burned in order to work the soil, leaving little trace of the forest that once existed. Thirty short years after this complete transformation of habitat, the grapevines were killed by an outbreak of vine aphids. The ranch owners abandoned the grape business and turned to sheep and then cattle raising, pushing the memory of the forest even further into the past.

This recent history is still visible on the now open landscape. The furrows of the plow line the meadows; grape stakes lie scattered in piles. Even the winery foundation is more or less intact. In the misty fog I can easily conjure up another time and place when this tree was surrounded by grapevines in hilly rows. It is much harder to go back further in time to visualize the tree's earlier companions. The empty slopes have been completely altered in vegetation since then. Grassland has replaced the vineyards and woodland; the few scattered trees grow only down by the spring or by the rocky headlands. I keep wanting to call this tree the Lonely Giant. It is almost a museum piece, an antique from a once much larger forest.

When classes come to the reserve to learn natural history, they sometimes visit this bay tree and hear the story of the changing landscape. The children walk the trail across the open savanna to the giant tree, drawn to it as a place marker of another era. Groups of twelve to twenty can easily clamber up the tree, filling the many different crotches of the six main trunks, which each support two or three spreading limbs. The tree makes a wonderful backdrop for a posed photo, beaming faces poking out from the deep green moss and rough bark. But it remains an oddity, a strange and unusually large tree, something to remember after you leave. The tree no longer functions as part of a forest system; it is a remnant, and what little is left of the forest speaks only through the imaginations of those who visit.

I feel some nostalgia for the former forest, wishing it were here for company. The short history of winemaking hardly seems worth the great loss of trees. I suppose the ranchers didn't expect to be put out of business by a voracious insect. But short-term decisions have consequences. As a neglected player in the business deal, the tree became isolated, its companions ignored. Without much choice in the matter, the pepperwood bay came to fill an unanticipated role as reference point in time and place, providing visitors like me a small glimpse into the history of the landscape.

CHAPTER 7

Magnetic Presence

IN 1964 A GREAT WALL OF FIRE BURNED FROM CAListoga to Santa Rosa, consuming fifty-five thousand acres and over one hundred homes. The flames sparked and leapt from tree to tree, exploding crowns in great starbursts, racing across the dry grasslands to the next wood in sight. The scope and power of the fire was fearsome; there was no stopping the rage on the landscape. The place was set to burn, and burn it did. The best human efforts only directed the fire; containing it was out of the question.

Down off Mark West Springs Road, the California Department of Forestry set a backfire going up from Leslie Creek to head the fire south instead of west. Like a small David rising up to meet the much larger Goliath, the backfire rushed up the hill, consuming coyote brush and chamise, scarring the trunks of venerable oaks and firs. The large fire, eating its way across the land, ran into the barren strip, hitting a wall of ashes. It turned south, drawn almost magnetically to more accessible fuels.

The fire was enormous. It consumed thousands of pounds of biomass. Chicken coops, town oak trees, vineyards, madrone groves all went up in flames in a matter of hours. It must have been terrifying for the locals who had hoped to keep fire out of the California landscape. This is impossible. The two are married to each other. Fire's spark has shaped this landscape and its inhabitants over thousands of years.

On one of the small ridges to the east of Leslie Creek, there is an odd swath cut through the vegetation, apparently a remnant of the firefighting efforts. To create a fuel break, a bulldozer had opened a fifty-foot-wide stretch, exposing the rock rubble and laying the ridge bare. In the last several decades the chaparral plants have grown back, covering over the disturbed soil layer. Deep red–trunked manzanitas now dominate the area, standing six to eight feet high, clumped close together in a huge bramble of crooked branches. There are two manzanita species here, both adapted for surviving the periodic sweep of fire. One type burns to the ground but retains a living core of root tissue that sprouts after the landscape has cooled. The other manzanita is killed outright by fire. The fire-adapted seeds, however, open in the 1000° heat and germinate in the ashen soil, taking advantage of the fresh deposit of nutrients.

Just beyond the bulldozed area is an elfin wood of large manzanitas. Under most circumstances a manzanita, with its bushy shape and low multiple branches, would not qualify as a tree. But in this spot the manzanitas grow twelve to eighteen feet high with canopy spreads as wide across as thirty feet. These Old Ones have so much presence that they belong in the spirit family of trees.

These unusual manzanitas, which by most standards should have been pruned back any number of times by fire, show no signs of fire history. Everything about them feels like an exception to the rule. It is as if time stopped and piled up in their trunks, turning them into sages of mystery. The sinewy, muscular branches are magnificent sculptures, inviting the senses, inviting intimacy. I have heard about two legendary giants in this area; one is the third largest manzanita on record, measuring seventy-two inches in circumference. This is eight to ten times the average manzanita girth.

I have walked perhaps a half mile along this narrow ridge, passing one stunning manzanita after another. These strangely undisturbed giants must be 250 to 400 years old. The feeling here is disorienting, like being in a peculiar zone where the magnetic field is reversed. Here in the land of fire, there has been no fire. This untouched ridge is an anomaly on the landscape, a place protected by the vagaries of circumstance and the direction of the wind.

Drawn into the thicket, I creep under the branches to the base of one of the giants. I gaze up at the smooth, red, twisted branches and small waxy leaves angled away from the harsh sun. The ropelike limbs carry currents of energy and water from the center core of the trunk. The branches divide and split—five, six times out to the pencil-wide tips—at each point taking a new turn. These manzanitas have not been beaten and blasted by the wind. The hard, slow-growing wood has taken its time in finding intricate direction and circuitous relationship. I feel pulled into the energy field of this vibrating sculpture.

Old Manzanita, I am quite dazzled by your electric presence. Shy and tongue-tied, I look away, not quite sure what to say. I see the Douglas fir seedlings scattered at your feet; I glimpse a squirrel slipping off in the distance. I want to speak with you, but you are an Old One. It isn't right to be too forward with one's elders. So, eyes lowered, I look around at your base. Tiny curls of deep red bark litter the ground, scattered among Douglas fir cones, oak catkins, and bits of lichen. Pale dry leaves of other years lie crumbled on the old soil, feeding the roots below.

When I first glanced closely at your thick base, the tears welled up. Look, the wrinkles. Elephant wrinkles, old people wrinkles, tree wrinkles. Above the corrugated base your trunk is smooth and unbroken. I touch your hard, cool skin; I cannot pull my hand away. I sense it is important to ask permission to be this close to you. You are one of the Old Ones. Next to you I am more like a mosquito than a rock. You have been here a long time; that alone commands respect. But even more, you have remained unscarred by fire and violence. You represent the untouched, the accidentally perfect, the rarity of growing to full maturity without major setbacks or devastating traumas.

Oh Old One! I want to know you. I am lying on my belly at your feet, my elbows sticking in the dirt. I want to speak with you and listen to your being. How do people and trees talk with each other? This is my question: what do you know? Crickets, mosquitoes, and the gentle breeze blowing over the lip of the ridge—do you feel these tickles? It is what I notice, living in a faster time frame. But you know something about years, lots of years—drought years, wet years, fire years. You are one of the Great Ones.

I want to explore every inch of your stunning architecture, the twisting, curving rivers of time molded into your sinewy branches. With my belly over your roots and my head on your shoulder, I want to get as close as I can to the core of your being. The pull is magnetic, inscrutable. I feel embraced by your presence; you hold the depth of time that lives inside me.

This great depth we share—it moves me to tears. Caught by your magnetic presence, I weep with the enormity of what is in each of us—full, rich, unfathomable. I weep for the sweet opportunity to be here, for the gratitude of simply meeting you. Amid the fire scars of history I taste the transience of life-forms. It might have been my friends and their homes in the fire of '64. To live in California is to expect to be burned. Landscapes turn over quickly; today's chaparral is a collage of fire stories layered one over another since the great drying-out time that began forty million years ago. I reel with the dizziness of so much change over time.

I pull myself up from the earthen floor and settle my back against your hundreds of years, seeking temporary stability. A big wind blows through, shaking all the shadows. But you are stable; you don't move. The wind goes around us; we stand still together. I begin to slip into your time zone. It is as if I've left all human memories behind. Here, in the energy field of the giants, I become manzanita, tree trunk, hobbit person low to the ground. I sit like a Still One. The wind rustles, and I move with the shadows, feeling manzanita energy passing through me. Ancient time energy, ridgetop energy, wind energy, the place-that-has-remained-wild-and-untouched energy.

If I stayed here long enough, it seems like I would grow into a manzanita. Put down roots and join back to back with you, Old One. That would be fine; some part of me is tree. I sense the core of your being that has been still for hundreds of years. This is how I know you; perhaps this is how you know me. You are a stillness container; so am I. I am small and young in the scale of your life, a fragile being in the history of time. Yet we meet in this vastness where nothing seems to move. This instant of remembering goes to the core of who I am, of where I've come from, and of how we are related by time and fire.

CHAPTER 8

Lifetime Lovers

I BALKED ABOUT COMING UP HERE. IT HAS BEEN over seven years since I climbed the hill to Cloud Mountain. Before I even started out, the journey was filled with memories, many of them disturbing. I lived on this Mendocino ridgetop for a year with a collective of college classmates and their families. Most of the seven adults and four children had been here for ten years, sweating together through various stages of homesteading—installing a windmill and water system, building temporary and then permanent housing, laying out a garden and orchard. They had baked bread, played music, smoked dope, partied, and squabbled through the turning seasons. They had grown up together on the land.

By the time I arrived, a year out of graduate school and discouraged with the job market, the community was crumbling. People were sick of eating with each other. They had lived communally long enough and now just wanted to be left alone. I, on the other hand, had not lived in a commune and was full of back-to-the-land fantasies. I was thrilled to stack cords of wood and read by lantern light. I savored the views from the outdoor shower. I was proud of my twenty foot by six foot compost pile. But I was lonely. Days went by without anyone reaching out to me. The three families cooked in their own private kitchens; only one woman and I occasionally crossed paths in the almost abandoned community kitchen. In the stark presence of the land I often felt naked and alone.

To compensate for the lack of human company, I sought out the trees. Douglas firs, redwoods, and madrones all grew well on this hill, two thousand feet above Anderson Valley and the town of Philo. I walked the logging roads to the nearest neighbors and down to the river, looking for new wildflowers, looking for anything that would keep my mind off the community's soured history. Without hardly trying I fell completely in love with the land. It nourished my spirit and soothed my heart at a time when the people around me were pulling away from each other, sometimes with strong words. One by one I shed my illusions about the joys and simplicity of living in the country. I didn't know it could be so isolated and difficult. I began to understand how people go crazy in remote areas.

I left Cloud Mountain after one too many complicated love affairs wracked the community. Even under the influence of homegrown sinsemilla, it was impossible to mask the suffering caused by multiple intimate relationships. The flower children's age of love had crumbled on the mountain. For me, however, leaving the land was the greatest pain. I was homesick for months, aching with the memory of the ridgetop in my body.

Away from the place I began writing and dreaming about trees. I remembered the tall tan oak and madrone on the edge of the ridge. I had spent many nights under their open branches, caressed by the wind and other lovers. I wanted to see these trees again, but I was afraid. It would be too much like seeing an old lover. I was not sure I could handle the intensity. But, ten years later, despite my reservations, the memory of the trees was powerful enough to draw me back to the land.

The road to Cloud Mountain winds steeply from the valley floor to two thousand feet over a twenty-mile drive. On top the ridge looks out over a verdant, rich riverland filled with lucrative vineyards. Like most country roads this one curves obligingly around the soft and hard slopes of the land, revealing the topography rather than overruling it. Thus one's pilgrimage is a genuine return to the land at every bend of the road.

Once I started up the mountain, there was no turning back. I knew this road like the back of my hand, and each grove of trees or shift in view marked a familiar sight. Here were the steep dirt slicks that were treacherous mudholes in winter, dangerous dust spinouts in summer. And here was the place the truck went into the ditch and got stuck for a week. The road was the first test for any visitor to Cloud Mountain. Guests weren't fully initiated until they had driven the road in a storm, dense fog, or snow. Then they could contribute a story to the mythology of those who knew the journey well.

I stop to unlock the first gate. Here is the point of request. How much do I want to speak with these trees? The invisible guardians of the gate prompt the question, reminding me of my intention. I stop again for a second gate a few miles later. Drat, a bit of trouble—the unfamiliar lock is stubborn in my hands. I can't get the combination to work. Will I have to walk the remaining five miles? I ask myself how much I am willing to do to see these trees. How many hurdles must I cross to return to this piece of my history? I try the numbers again as doubt worms its way into my fingers. In spite of my tension the lock finally works on the sixth try.

To reach the third and final gate, I drop over the lip of the hill. I am poised on the edge of entry. What will I find ahead? How will the trees look? In the midst of so much logging in this county, will the trees even be standing? In the last seven years all the families have moved out; only one person is left from the original group. My classmate who had bought the land and started the community died of lung cancer shortly after the birth of her third child. The shock of her death frayed the last threads of her utopian vision, scattering the residents their separate ways. So much suffering. I return a little sobered by the tragedy of human lives.

But the trees—will they still be there? I feel some anxiety, like a lover returning to the beloved. Will these two trees I have loved still be among my living relations? This particular anxiety is haunting and familiar. I carry this question with me constantly—will the home planet I love so deeply continue to exist? The anxiety tightens my stomach into a knot of hesitation. I don't know if the trees are intact; I only know this haunting, anxious feeling.

I enter the last gate and circle the edge of the forest to the open grassland. The familiarity of this view comes rushing back with surprising power. The mountain drops away almost half a mile straight down to the lush valley. Row after row of gentle peaks stretch to the south and north; fog hovers over the ocean. I walk slowly to the knoll. Here are my beloved trees. I am so relieved to see their majestic forms. Tan oak and madrone—these two trees hold up the sky right here; no others are as tall. They own the land around them; their roots fill the entire area.

The madrone commands a certain presence, with its distinctive central fire scar and unusual size. The base of the trunk is easily five feet across, supporting three large branches, each over two feet in diameter. The tree leans to one side, compensating for its asymmetry with balance and poise. This is quintessential madrone—warm red trunk, muscular and smooth, curled parchment peels crunching underfoot, dancing branches and leaves. I greet the tree with my hands, letting them find the memory of madrone in my fingertips.

The first time I fell in love was in a grove of madrones in southern Oregon. I had gone camping on Mount Ashland in the Siskiyou National Forest with a new friend, who had begun to capture my heart. We spent a remarkable night gazing into the arms of these lovely trees. We rolled and tumbled alongside the sinewy roots, kissing the dark night and watching the moon as it shouted through the leaning branches. We played like fox cubs, whispering secrets to the laughing trees. I was completely enchanted—but by whom? It's hard to say whether I fell more in love with the young man or with the madrones.

Each encounter with a madrone rekindles my original love affair with them. I look at this handsome tree and feel my heart full with every madrone I have ever loved. I feel the sweetness of reaching out to touch the muscular limbs, the leaning branches, the solid trunk. This is a delight I can never forget. Always I yearn for the physical remembering. This is why I came—to rekindle relationship in a specific time and place, as one physical being to another.

One new thing I notice are tire marks in front of the tree. One of the residents must be driving to her home instead of walking. The road seems to have cut a path of unconsciousness between the trees. This spot had been a sacred grove for me, a peaceful sleeping place under the starry sky. The evidence of regular vehicle use erases some of the stillness I remember.

I walk down the road to find the last resident of the old community. Her little boy comes out to greet me. He shows me all his toys and books while I catch up on history with his mother. Among other stories she tells me that the tan oak lost its top a year ago in a late winter storm. The whole crown crashed to the ground, scattering branches everywhere. Probably the madrone was the only witness to the traumatic event.

I walk back to the two trees and lean against the tan oak, grateful that it survived the smashing blow. The tree seems healthy in spite of its injury; the only signs of trauma are a few remaining piles of dead twigs and leaves. The lower trunk is solid and the outer branches thick with leaves and catkins. I believe this tree may last many more years in this spot.

Either tree alone would be quite commanding, but with the other present they tell another story. I can see where the roots and branches mingle, establishing a conversation in the soil and air between them. The energy passes between these trees as they stand in relation to the same physical forces on the knoll. Together they shape the passage of the wind and its sound between them. The place has become the expression of their lives.

It is humbling and quieting to sit between these two trees. I remember the painful isolation that lured me to them years ago. I was dependent on their friendship for emotional survival, since the human community was an uncertain source of love and stability. These trees, however, offered their open arms whenever I came to visit. I sense in them a story of age and weathering, the cumulative richness of maturity. These are not young trees starting out fresh against the elements. The oak and madrone have withstood the ravages of cold, heat, wind, fire, roads, and insects. And they are still here. This is a mark of ecological wisdom, the intelligence of survival, the simple act of living through it all.

Now I see what drew me back here—the need to remember what it is to just survive, to exist, to go through the bare experience of living through whatever comes. These two have done this together in the same space and time frame. This is not insignificant. Their knowledge of age and relationship is quite tangible; it is what trees have always offered. But now, ten years down the road, I recognize this wisdom in a way I never knew before.

CHAPTER 9

Mystery Pine

Four days back from Thailand, and my first day on campus. The semester had barely started when I left; now I was doing my best to reorient to the weekly rhythm of teaching and learning. I had traveled to Southeast Asia to lead a training for Buddhist monks and social activists working on human rights and social justice issues. Our Thai hosts had arranged for a small delegation of concerned Buddhists to travel to the Burmese border to meet some of the monks, university students, and indigenous people who were seeking refuge from the repressive Burmese government. We rumbled through the opium hills by truck and listened to the stories of the Pa'O people guarding their border camp. We met young students from Rangoon who had learned to live in the jungle and were now teaching the local Shan children in a makeshift forest school. The most disturbing accounts came from Buddhist monks, whose monasteries had been surrounded by tanks and closed down to limit their protests. Over two thousand monks and students had been shot since 1988, when the military regime refused to turn over the government to the democratically elected Aung San Suu Kyi.

I wanted to know about the teak forests in the northern region of Thailand near the Burmese border. Logging had accelerated in the late 1970s, with large concessions going to Japan, Sweden, and the United States. In the short period of a decade over seventy percent of Thailand's tropical forests had been cut, and the country had suffered massive flooding and devastation. As a result, the Thai government imposed a moratorium on logging. However, they continued to trade and cut teak at rapacious rates.

Our Thai hosts showed us some of the young teak forests that had been replanted for future harvesting. Teak grows very slowly, so it would be at least a hundred years before these trees were of much commercial value. To correct a serious post-deforestation cash flow, the Thai government planted thousands of acres of fast-growing trees. In some places the native forest had been deliberately cut and replaced with nonnative pines and eucalyptus because they were more favored as trade species. The pine plantations looked strangely out of place and sterile. The original forest had been leveled and burned, leaving no trace of the former complexity of vines, shrubs, and canopy trees. In its place stood rows of evenly spaced pines from another continent, obliterating the intelligence of what had once grown there for thousands of years.

My return to campus was a return to rhythm and obligation as well as to home base. I tried to find the old time zone in my body and to make sense of my impressions of Southeast Asia. But I found myself resisting the pressure of the familiar. The rapid, unmonitored loss of northern forests was deeply disturbing. I needed time to wander and contemplate the incongruities of what I'd seen.

So I walked. I followed the south fork of Strawberry Creek to a corridor of coast live oaks adjacent to a fenced-off construction site. Too familiar, I thought, both the trees and the construction. I wanted to speak with a stranger. I wanted to meet someone exotic, a foreigner. I myself was a foreigner returning to my own culture. I was looking for something unfamiliar and unknown.

The Berkeley campus had once been a spacious, park-like setting with well-defined open spaces between buildings. The founding fathers had planted trees to enhance the character of the learning landscape. Now, over a century later, the trees have matured to a stately grace and grown into the vision of their planters. Some, like the dawn redwood, retrieved during the first U.S. plant expeditions to China in the 1940s, are botanical treasures. Others are simply horticultural conveniences, trees that frame classrooms or cover utility pipes. Both, however, require the careful work of groundspeople and arborists who prune them in winter, water them in summer.

As the campus grew in size and student population, the buildings encroached upon the space occupied by trees. In sheer volume the built architecture dominates the university. But the trees are key to the shape of the landscape. Next to the ponderous and thick buildings, the trees suggest motion, growth, and quiet spots for thinking.

I wandered across campus. Up past the north fork of Strawberry Creek, I found a curving promenade of ten exotic pines. The trees were well placed, thirty to forty feet apart, leaving ample space for each individual canopy. As a group they formed a conspicuous element of the landscape design. I moved from bench to bench, trying out several perspectives. The trees were too far away. How can one be intimate at fifty feet? From the benches the trees were distant objects on the horizon. I was seeking a conversation; I needed to meet these strangers up close.

I found a well-worn sitting spot at the base of one of the trees. Others before me had spent time here between classes doing schoolwork, talking with friends, or dozing in the warm sun. I knew little about these trees except that they were not native to this region. Perhaps these are the pines they are planting in Thailand. It seems awkward not knowing the names of these trees. I feel ignorant, perhaps even rude. One would not sit so close to a human stranger without being properly introduced.

Despite their foreign origin, the immigrants have adapted well to this climate. Their furrowed trunks support healthy crowns of green needles. But, of course, the trees have been cared for. On the Berkeley campus, people are paid to look after the trees, to trim back the heavy branches and feed the soil. Professionals have asked, "How are you? What do you need?" They have tested the soil and loosened the grass, collected the fallen pine boughs, and checked for disease. I can feel that history in the presence of these pines; their wildness includes some experience with people. Activities like this leave a trace of kindness around a tree.

As exotic trees planted on California soil, the pines are displaced from their native land just like the pines in Thailand. This word *exotic* conjures up a sense of adventure and travel—like the foreign land I have journeyed to. A curious botanist probably wondered if the strange and beautiful pines he saw growing in a foreign land would be able to grow in his native country. He brought home what was fascinating and different and domesticated it to become part of his world. This proclivity for the exotic is at the heart of horticultural adventure. The urge to domesticate, improve, manage, and manipulate easily grows out of the first desire to colonize the unknown. And with only a few more steps, we have tree plantations of strangers, completely out of place, altering the wild character of the landscape.

As I come closer to meet the tree, I experience a tension of boundary. To greet the mystery of the Other requires a conscious taming of fear, a willingness to be present despite the barriers of difference. It is easy to follow distracting detours away from the actual meeting place of mystery. But I did not come wandering to be distracted or entertained by my own thoughts. I came to meet a foreigner, to remember my own foreignness. I reach out and touch the tree. Here is the actual mystery—the contact zone of questioner and question. These deep fissures in the pine bark—how did they form over how many years? These mini-ledges in the cracks—do nuthatches and brown creepers use them as snack zones? What beetles and crickets live in the layers of bark? Why are the sprays of long needles restricted to the tips of branches?

A scattering of ants trickle down the trunk to the ground, picking their way through the rubble of bark. The hard dirt reflects the recent dry years and compaction from visitors sitting in this spot. The lack of invertebrate diversity is noticeable. No crickets, no pillbugs, no millipedes or centipedes, not even much evidence of moles or gophers. I suspect this soil has suffered from the impact of chemical control. Like the pine plantation, this landscape is sterile, despite its beautiful setting.

Beneath the grassy lawn is a subterranean tangle of roots. I imagine this territory to be chaotic and unruly, the roots growing every which way. Like most in my Western culture, I have been trained to equate the dark and the unknown with chaos. But just because I can't see what the tree is doing, is that any reason to assume the tree does not know what it is doing? The roots go where they find passage; they make their own way through the orderly routes of least resistance—along rocks and hard places out to the softer spots. Each turn of the roots is a statement of design, a pattern of history. The growing tips seek out the areas of greatest moisture and softest soil, curving through the substrate like water through a creekbed. The design is of choice and circumstance.

I want to go down into the dark and know the labyrinth of support and nourishment. I want to go down in the burrows of beetles, through the tunnels of worms into the maze of root hairs pushing through the dirt. I want to go down into the roots and know the origin of this tree—what is its homeland? I want to go down into the mind of chaos and meet the mind of control. This pine and the pines in Thailand have a common story—the simplification and sterilization of mystery. The mind of chaos is the mind of wilderness—pure expression of uncontrolled and untameable life force. The mind of control is the mind of plantation—in which trees from a strange land are slaves, planted to serve their masters. In the dimming light I begin to see the pattern of strangeness on the land.

CHAPTER 10

The Golden Time

NOW THE DAYS ARE GROWING SHORTER AND THE sky is speaking of winter. Change hangs in the air with a question mark. The harvest is in, and the season of waiting begins. This is the tender time of year, when everything drops away. The trees are left bare and empty, only skeletons of continuity remain. It is the tender time because the impermanence of things is so visible, so unavoidable. The winter heart feels the lengthening darkness and the turning of the sun.

I am noticing this turning in the company of a magnificent ginkgo. With a canopy spread of close to a hundred feet, this tree is one of the most commanding individuals on the U.C. Berkeley campus. Its bright fall color lights up the space between the dark green redwoods and the pale granite building. The ginkgo is a campus landmark, well loved by those who watch it change each autumn. In the last few hours of afternoon light the ginkgo is sunbathing in the warm, golden rays. The leaves have begun to say good-bye, fading from green to gold. The whole tree shimmers with yellow. Students walk by on the path, lost in equations and other philosophies. The light is an event above their heads. Few stop to look up or notice.

For a month I have been anticipating this golden time. Last fall I taught an evening class in the nearby building, and every week we watched the ginkgo slip closer to winter. By the end of the course, the tree had shown us its green finery, golden halo, and finally its elegant skeleton. The course was a botany class for garden volunteers, charged with communicating their love of plants to visitors. As a representative of the oldest living tree species, the ginkgo was an inspirational teaching companion. Though they no longer exist in the wild, they are favorites among horticulturists for their glorious color and elegant fan-shaped leaves. The 150-million-year lineage of ginkgo ancestry exists today only because the Chinese preserved these trees in ancient sacred groves.

Now, in the closing days of November, I am drawn to the ginkgo again for its poignant display of impermanence. I have come here in the cold, damp afternoon to seek out the warmth of an old friend. I can feel the season of stripping away coming on; I need to prepare for the dark heart of winter. I have come to be with this ginkgo especially because of its capacity for beauty in the midst of radical

change. I would like to know how to do this—how to gracefully lose everything on the surface and yet retain a core of life and vitality on the inside. This is a fine art, and the ginkgo has perfected it. The elegant leaves will drop all at once in a week, becoming a pool of gold at the base of the trunk, the last gift of sunshine to feed the tree over winter. The tree will rest naked in the dark time, drinking up the last tastes of gold through its subterranean roots. This would seem to be excellent practice for preparing to die—letting go of form, letting go and yet radiating with the richness of life's light.

As the light drops in the sky, the color changes from cool yellow to brilliant gold, catching the horizontal light. I can almost feel the warmth from the fire in the tree. It seems as though it would glow on in the dark even after the sun is gone. I've been told that the exquisite fan-shaped leaves symbolize the qualities of stability and impermanence celebrated in the Japanese art of tea. In this elemental and highly refined art form, one offers a cup of tea in recognition of the always passing beauty of life.

As the light grows dim, the ginkgo enters the long night of winter—the going-under time, the black cloak time, the time for burrowing deep in the ground. The tree has been nourished by the spring sugars and now consolidates its gains in preparation for the cold, short days. Winter extremes are stressful; it is not easy for any plant or animal to survive cold spells, although some do so better than others. If you are a tree, there is nowhere to go to get warm. These brief sunbaths are at least a temporary respite from the grip of harsh weather.

The light is sinking low to the horizon and slipping off the lower branches. Cold shadows creep up the tree; activity will soon be slowing down. But the upper branches are still full of delight. The corners of the mouth are smiling. The tree knows this yearly rhythm in its own way, without words. It knows about time and relationship, and the slow, undulating movements of planets and stars around golden trees and tender hearts.

Standing here is a way of marking slow time, of putting my feet in the context of years instead of minutes. I need this slow time for a proper frame of reference on the year's activities. It is a help in settling down for the winter, for the deep time. This is the time of cultivating patience, of nourishing the soul, of soaking the ground thoroughly in preparation for spring. I don't know which seeds beneath the surface will germinate. I can only take care of the ground of darkness and sit still without waiting. The deeper the stillness, the greater the capacity to respond to change when it comes. I go into the darkness with nothing tangible to guide me. I have only my willingness to be here through the deepening.

Now the tree is half in shadow as the day passes into dusk. The sun almost flat on the horizon announces the impending good-bye. Such drama before parting! The lowest and last light is the most intense, almost blinding to the eye. The poignancy of impermanence comes again—this tenderness, this nearly imperceptible change that keeps revealing itself to the watcher. When the light decreases, the spiral turns inward. Then it is the inner light that keeps burning, that serves to guide in the darkness. The dark of the year is irrefutable. It can be obliterated by electricity, but still it comes. The great darkness always comes, creeping up the ginkgo, coming around again for another year.

The great darkness is the story of how people and trees come through the testing time. The tree will be tested by storms and cold and heavy winds. It may lose a branch or two or catch a cold. If there is another dry year, the tree may lose the flexibility it needs for winter. I imagine the pool of golden leaves feeding the golden circle within, keeping the inner tree alive through another hard season. And how will I come through the testing time? I don't know; I cannot know what gifts the darkness will bring.

The sun has set, and the top of the tree is drifting into silence. Streetlights have come on along the path; the sunbath is over. The turning point of the day is not unlike the turning point of the year. Both are delicate transitions from one extreme to another, fraught with the danger of hurrying. The voice of impermanence is the only one singing into the black night, with little to offer the listener. In the poignancy of this melody it is hard to leave the light. I am hesitant and uncertain in welcoming the dark. Bound together in this rhythm, the ginkgo and I walk into the night.

CHAPTER 11

A Way of Looking

THIS MORNING I ROSE EARLY TO SPEAK WITH A tree that didn't know it was going to die today. The tree was a backyard elm that shaded the southwestern corner of my mother's new house in Portland, Oregon. Planted forty years ago when the development went in, this suburban tree was not part of any fragmented forest. It was planted as a horticultural decoration and visual barrier between neighbors' yards. With drooping, spreading branches, the tree dominated the property and acted as guardian for the backyard. Its shapely limbs offered a modicum of grace to this small section of the neighborhood. The tree, however, had unknowingly overstepped its bounds. With vigor and the force of life, the roots had begun to press up against the foundation of the house. In addition, the elm was a "messy tree," dropping thousands of leaves and small branches each fall, littering the ground with troublesome debris. Human inconvenience and fear of falling limbs determined the tree's fate.

When I heard the tree was slated for removal, I asked my mother to delay the felling until I arrived from California. Though I felt helpless to stop it, I wanted to be present for the deed. For almost a week before, the roads had been covered with ice and temperatures remained in the teens, making tree work too risky. But today, with a break in the weather, the contractors were ready to take the tree down. The warmer air and dripping rain had melted most of the snow. In their opinion this was hospitable weather for tree felling. They thought they should take advantage of the break before the next storm hit.

This week my family had come together from London, St. Louis, and California to celebrate Christmas in Oregon, our forested homeland. My two youngest brothers had strong feelings for trees, and I thought they might join me in this tree event out of arboreal devotion. I had a vague idea about holding a family ceremony for the tree as an attempt to address the loss together. I thought we could serve as witnesses for its demise. Would the offer of solidarity make it any easier for the tree to die? I didn't know.

When my youngest brother was nine and I was away at college, the highway department laid a new freeway through the deep ravine behind our first Oregon home. They cut and felled a wide swath of Douglas firs as the price of connecting two major highways. My brother was

very upset about it. He composed songs to the trees and wailed his loss on the piano. He wrote detailed letters to me chronicling each stage of destruction, calling for company in his anguish. He was having a tree awakening, and I was his witness. Each new phase of clearing brought another round of sorrow and protest. Of all the children in the family, he was the only one to have grown up in that neighborhood since birth. He knew every bike and footpath through the trees; the woods were his playground and place of escape. He was heartsick watching them disappear.

While he moaned for the Douglas firs, I fretted over five giant sequoias planted in a grove at the end of the ravine. Taller than the firs and distinctive in their arrangement, they stood out as beacons from several approaching roads. Their perfect conical tops were a reference point for me, a sign of the home landscape on my return from school and other excursions. As a child I had often walked under their spreading boughs in search of solitude. Now the road-flagging tape was laid dangerously close to these five prized sequoias. By the time the asphalt was poured, two of the trees had died from stress and invasion, and I joined my brother in his mourning.

The second youngest of my four brothers was in love with the Douglas firs on the edge of the ravine. He had built a succession of platforms and tree forts in the high branches where he would practice his French horn in true Wagnerian style. These were his personal retreat areas, spare and simple in construction. Safety was a minor issue next to his passion for being with the trees. Despite one forty-foot fall, he couldn't be kept out of the trees, no matter how loud my mother's protests. For all three of us the woods in the ravine were a place to wander and to evoke the wilderness in our souls.

Now we were together for the winter holidays in my mother's home, facing the loss of the elm. Experts had been consulted, the matter discussed, a decision made. The tree would come down. I stood outside in the gray dawn light and gazed at the tree, rain falling gently on my face and shoulders. What could I possibly do? I felt obliged to mark this tree's passage from life to death; something inside was crying for the tree. My brothers stayed inside, leaving me alone with my uncertainty.

It was almost 8 a.m.; the tree crew would arrive soon. I looked through my suitcase for miscellaneous tools of ceremony. What is the proper ritual for a tree death? I didn't know. I had never seen one done before. Tree removal is something arborists do, a specialty trade like being a mortician. Most people don't pay close attention to this kind of work. Today I would. The sadness was gaining momentum. I thought a ceremony might at least deflect my grief.

With three sounds from a small bell, I began. I walked slowly around the tree nine times, breathing deeply, calling the tree people to listen. I knew almost nothing of the tree's history. I had barely begun a conversation with this tree. It felt like giving communion to a total stranger before death. Last rites, these were last rites. I lit a small candle and offered four sticks of incense to the four directions, placing them at the base of the tree. The tree was the centerpiece of its own altar, the altar of its death.

The tree workers arrived with their gear—chain saws, ropes, belts, and harnesses. They had heard I wanted to do a ceremony—what were they imagining? They pulled on their climbing boots and checked the chain saws. Only a few minutes remained before the first cut. It looked like they were preparing for surgery, only this was not going to be a healing operation. My brother used the word *murder.* Yes, why not call it that? It was a premeditated choice to destroy another living being. I chanted a dedication under the dripping rain, a request for forgiveness for those who plant trees too close to homes. I asked for compassion for those who are uncertain about how to care for tree beings and for those who suffer the consequences of loss of tree friends. I felt unprepared with tree prayers. I had never learned anything of this sort in Sunday school.

Three last bells and the short ceremony was over. It was a quiet act of intention that did little to reverse the fate of the tree. But at least the elm did not die alone. I brought the lit candle inside to symbolize the life of the tree. We could watch it burn down as the tree was dismantled. My nieces and nephews were awake now and wanted to know what I was doing. My brother thought it better for them not to participate in the ceremony. They wouldn't understand, he said. They would ask, "Why are you killing the tree?" Children are too young to understand, he said. I wondered.

Maybe children are the *only* ones who understand. It made sense to them that the candle was the tree.

Chunk by chunk the tree workers handed down the limbs of the tree. They worked carefully and skillfully, drawing on years of experience with tree morphology. Even in the rain they placed the branches precisely, never hitting the roof or fence during the three-hour process. When the last big section of trunk fell to the ground, it shook the house with a solemn thump. My nephew came running in. "Aunt Steph, Aunt Steph, the candle fell over!" The children explained the obvious: "Of course," they said. "The candle fell over because the tree died."

I had thought the ceremony would be a family event, but in the end I was alone with the tree and the dilemma of death. Maybe the candle was the most important piece of the story. The children understood the life of the tree going out like the candle. They could see as well as anyone that the tree was dead. I wondered if I would have the courage to be honest with them about my feelings. I felt self-conscious and culturally inadequate. It is not common for Americans to consider that trees carry spiritual value. I wanted the children to see how our choices as human beings affect the lives of others, but I didn't know how to talk about it. In my tongue-tied empathy with the tree, all I could do was watch and stay with its spirit presence.

As the workers sawed up the corpse into firewood, I held my brother's baby by the dining room window. His eyes never wandered from the activity. We were both paying close attention to what the men were doing. What *were* they doing? A baby has a way of looking that makes you look too. "Yes, little baby, here are the people and here is the tree. The tree is on the ground now, laid flat. The people are taking the tree apart. The tree is dead now. The tree is a pile of wood."

Does a baby understand death? I wondered. Or does he sense my concern and helplessness? I was grateful for his company in this time of watching. There was now a hole in the universe where the tree used to be. In a few rainy hours the architecture of graceful branches and arching limbs had been reduced to a foot-high stump surrounded by sawdust.

My mother had consulted an expert, and the expert told her what to do. It will be better for the house, he said. It is something you should do. The expert knew something about trees that she didn't know, and because of her lack of knowledge she felt obliged to defer to other voices. But I had wanted to speak with another voice, the voice that speaks from relationship with trees. Though I did not know what to say and my words did not alter the fate of the tree, I at least wanted another voice to be heard. I wanted to show the children there might be more than one way to approach trees.

Someone planted this backyard elm just after World War II, two generations before this baby was born. Perhaps that person expected children to play under its canopy in the warm days of summer. I found myself wondering what trees would be here for this baby as he grew up. It seemed difficult and clumsy to think into the future, imagining specific relationships between children and trees. Seeds of these relationships planted now might mature long after my death. I deeply wish the next generation a rich and considered relationship with trees. But how will children learn to care for trees? To be respectful? To pay attention to the old ones? Who will encourage them to cultivate depth and integrity in relationships with trees? How will they find their own voices that speak with trees?

A baby learns about trees by being around others who know about trees. Children learn from how they see people act toward trees at home, at school, and in the community. What are today's children learning from their elders? To look at the record of worldwide tree loss, it would seem the primary message is about consumption of trees for products. I cannot accept this as adequate. It cuts too forcefully into the generative core of life. The children will ask later, was this much destruction necessary? I want to have an answer for them. It will not do to simply pass on the old ways without question.

PART THREE

Entering the Tangle

CHAPTER 12

Fallen Tree

BLUP-BLUP-BLUP-BLUP-BLUP-BLUP-BLUP-BLUP-BLUP-whrrreeen, whrrreeen. My peaceful day at home is interrupted by the biting sounds of a chain saw. *WhrrrrEEEeeeeeEEeen.* Someone in the neighborhood is cutting up firewood, I suppose. *Blup-blup-blup-blup-blup-hummmm-whrrrrEEEE-whrrr.* From my deck I gaze out across Muir Beach to the Pacific Ocean. Ahh! Such a lovely sight in the crisp air of early winter. The thermal currents have begun to shift, clearing the fog and smog from the autumn sky. *Blup-blup-blup-blup-blup-blup-blup-blup.* The bright ocean waves are crashing—*whrreEEEeen*—in a perfect surf roll. *Hummmm-blup-blup-blup-blup-hmmmwhrr.* Early returning Monarch butterflies drift in the sunlight outside the house.

Despite my attempts to concentrate on the glory of the day and the humble miracle of migrating lepidopterans, the chain saws are rather hard to ignore. *WhrrEEEEEeeeee-EEEnnnn.* It has been my accidental good fortune to share this mini banana belt where I live with a large cluster of butterflies. The Monterey pines on this knoll are protected from coastal storm winds and chilly creek bottom air. The butterflies have chosen the nearby pines for their local overwintering spot; the first several hundred are settling in this week. Last year three to four thousand Monarchs clung to these trees in massive orange clumps, seeking respite from colder climes up north. From Mendocino to Carmel, hundreds of butterfly trees host thick swarms of orange-and-black travelers each year. The butterflies had abandoned the Muir Beach site for several years. But now they are back, and I am eager to have these first returnees feel welcome.

The chain saws, however, are not helping one bit. *Whree-EEEeeeEEnn.* I am drawn off the deck to investigate. The invasive whirs and whines pierce the air. I remember that one of the pines fell across our narrow, potholed road last week, blocking access for cars. Members of the volunteer fire department cut the trunk and moved it out of the way, leaving a craggy heap of wood on the side of the road. Apparently they've returned to finish the job and clean up the mess. Well, that's a relief. At least they're not taking down a living tree. At least the butterfly trees are not in danger.

The fallen tree must have been rotten at the base. Crusty with cones and lichens, this pine was certainly no more than 100 to 150 years old. Monterey pines grow fast and vigorously in their youth, but by old age they often become craggy and overweighted. This one, like a number of others along the road, was past its prime. The combination of drought and bark beetles had probably weakened it to the point of collapse. A large tree, it had once supported an active tree fort. Now the broken platform stuck up from the wreckage at an odd angle.

Looking at the tree and the men at work on it, I feel sad about the loss of the tree. But then, it died a natural death and it was fairly old after all, so I cannot be angry that its life was cut short. Though I hover in some middle realm between reason and passion, I accept the death of this individual tree. But I find myself wondering what the tree was doing here in the first place. Did it actually belong here? Botanists claim that Monterey pines are native only to the Monterey peninsula as far north as Año Nuevo, seventy-five miles south of here. Planted by early settlers to the area, this tree could technically be called an invader. Does this give it less of a right to exist? In many cases invasive plants have taken over the native flora, displacing or suppressing local biodiversity. Some of my neighbors have been concerned that there are already too many Monterey pines in the area. With this reasoning, should I rejoice at its death? On the other hand, cone fossils found in Marin County indicate the presence of earlier maritime pines in this section of coast. One could argue that the current northward expansion of Monterey pines is a natural evolutionary trend toward reestablishment of a similar type of tree.

It is difficult to sort out these responses without oversimplifying the situation. How does one relate to a whole species and its history? Individual trees offer the opportunity for one-on-one relationship. But a species is an idea, a cumulative history of countless individuals that have lived many millions of lives under varied weather conditions, changes in sea level, diverse soils and substrates. Can a person truly have an emotional response to such a collection of life histories? I don't know whether Monterey pines belong here. Yet still I must consider the ethical questions and make choices.

Meanwhile, the tree is coming apart. *WhrrrEEEEeee EEEEn*. The stream of sound slices through the limbs of this former resident. Native or nonnative, it is being broken into firewood rounds. The fast-working saws have completed the gross-scale dissection, leaving bits and pieces of the body strewn on either side of the road. Is it dead? Wasn't a large part of the tree already "dead wood"? The old pine is no longer standing vertically and is no longer connected to its roots. Yet the branches are green and apparently alive, still drawing nourishment from the stored sugars, still producing food from the sweet light of the morning sun. A tree is such a massive storehouse of energy and water, it can support life at its fallen branch tips for a long while. But not these branch tips; they are being clipped off at a rapid rate. From a wood gatherer's point of view these green branches are the least useful part of the tree. They are stripped away quickly to get to the wood, clearing the living from the dead to accelerate the dying process.

If the tree had fallen in a less inconvenient place, perhaps it would have been left to die undisturbed. The rough and springy branches would drop off one at a time as they dried out or rotted, leaving a random scattering of curving twigs. The green needles would photosynthesize until they ran out of nutrients and turned yellow, then brown, falling off the stems after a few weeks. The tree would fragment into smaller sections of digestible territory, and these would be colonized by termites and bark beetles, grubs and worms. The grassroots power of insects would erode the bark and wood through several generations of chewing. Fungi would invade through the tunnels bored by the beetles, consuming the soft heartwood. Pine-loving boletus mushrooms would penetrate the surrounding soil, weaving a lacy basket of mycorrhizal roots to catch the decaying nutrients. This long, slow settling in of decomposers would be a much more gradual process, much quieter and gentler than this chain saw funeral.

But in this situation there is no time or space for the luxury of a fully lived death. The pungent, refined smell of gasoline fills the air; the biting sound revs up again. *Whrrr-EEEEEeeenEEEnn.* It is so insistent, so aggressive, so capable of overriding the resistance of the tree. The main trunk has been severed into four large chunks, and all the side branches lopped off. The tree is beginning to look more like a woodpile than a pine. I am beginning to feel worn down myself by the relentless growls and snarls of the chain saws. The penetrating vibration presses into my chest, biting away at my heart, carving my trunk into segments. *WhrrEEEeenEEnn.* The racing sound slices through tree and neighborhood, dominating the quiet butterflies and the roaring ocean with its presence.

The tree is quickly being reduced to unrecognizable pieces. I can see this loss of integrity, this breakdown of a system. An organism has a history and a shape to it; these don't disappear immediately upon death. In a more natural death, the shape changes gradually as parts decompose, and the history is gradually forgotten and replaced by more history. I find it easier to accept the gradual process; it is more familiar to my evolutionary experience of the natural world. The chain saw's acceleration of this transformation of life is jarring and full of hurry; it tenses my stomach and leaves me dizzy.

I have held a chain saw before and cut wood for my own winter fires. A power saw has strong energy and can be a dangerous tool. The tiny teeth bite over and over again at an invisibly rapid rate. The saw gulps its way through the wood in split seconds, consuming its meal voraciously. There is barely a moment to see what is going on. The tool is more powerful than the hand that holds it. It is powerful because it is so fast, dangerous because it is so forgetful.

I, too, know something about moving fast. The faster I move, the more likely I am to forget where I am. There isn't time to notice and be present with what is right in front of me; I am too busy moving past it. The faster I go, the more I consume unconsciously. I eat the ground below my feet without even noticing. I become lost in the dizzying thrill of speed. I forget where I am because I am cut off at the roots. It is that root connection that maintains the life of the tree and my life. It is not possible to stay rooted at breakneck speeds. Can a chain saw stay conscious at such speeds? I don't think so. And thus it breeds unconsciousness in its users.

The broken tree has collapsed into a jumble of branches, logs, needles, lichen, and bark, surrounded by unconsciousness. Ragged breaks reveal the fresh, bright inner wood, rough around the edges of its wounds. The corpse has been shredded with the skill of the fragmenting saw; it is a sad burial for a member of the neighborhood. During a sudden pause in the operation, I hear the ocean again. The workers must have run out of fuel. I wonder if the tree fragments can hear the ocean too. The pine must be in a state of shock after the intensity of these electrical vibrations. It is enough to shatter anybody's peace of mind. I believe the tree must have had some kind of peace of mind before it fell and the wrecking crew arrived. Now everything is a jagged heap of branchy spikes and swords, less dangerous for cars, but also less beautiful, for the thing no longer has a shape of its own. It has become a junk pile of unconsciousness.

Someone will use this wood for firewood. But what will be burned in the fire? The wood, the tree, or the unconsciousness? Will anyone remember the nature of the tree that became the wood? Perhaps. More likely the spell of forgetfulness will carry over to the next person who handles the pieces of wood. The logs will burst into flames and become pure energy, energy purified. Noticed or not, this will be the final cremation for the fallen tree.

CHAPTER 13

House of Wood

I HAVE WOKEN UP AT THE END OF A LONG WEEK OF tiredness. I am too tired to go anywhere. Too tired to seek out a tree for comfort. Too tired to walk in the forest on the mountain. Too full of the sadness and tenderness that speak through me as I teach about how we are living with the environment, how we are dying with the environment. It is difficult work to be present with the state of the world. The more I pay attention to the economic and political forces driving environmental deterioration, the less certain I am that anything I do will stop it. My heart aches for the thoughtless deaths of so many trees. Sometimes I long for a break from the destruction and grief.

Here in my home I find some comfort in the beauty and simplicity of this house. I am grateful to be surrounded by wood and by the memory of trees. Wood walls and ceiling, a beautiful oak floor, paned glass and wood windows, kitchen cupboards crafted of wood. From all sides I am embraced by wood. The presence of trees soothes my eyes and soul. The natural warm brown color is restful. It is just what it is, nothing extra. No decorations, no wallpaper, no paint, no layers of anything masking the wood. The simplicity is refreshing. I appreciate the unevenness and random variation of the wood.

All these trees—the oaks in the floor, the firs and redwoods in the walls, the cedar in the yarn chest—are trees of the Pacific forest, trees of my homeland. But here in the house they are quiet and alone, no longer dancing in the wind or singing with the birds. It feels a bit like a tree cemetery—in elegant form, of course. It is hard to think of the wood as dead. It doesn't feel like I live in a house of death. The grain of the wood is too alive. Its memory is too vivid, etched from the experience of lifetimes. I feel the histories of individual trees; they resonate in each beam and board.

One thing is wrong though—the straightness. All of the wood has been cut into straight forms. Trees, however, are not entirely straight, especially the hardwoods. It is convenient to live in this straightness. It makes walking and organizing things easier. It works well with gravity and the desire of the inner ear for balance. But I miss the graceful curves of the living tree. I miss the tangle of branches, the intimate spaces between the twigs and fingers of each limb. Planed surfaces in a house have all the intimacy ironed out

of them. They have been flattened, standardized, regulated, cut to conform to human design. In the process the trees' own naturally beautiful shapes have been altered beyond recognition.

So this is the pain of it: in leaving its life-form behind, the wood has become an object for human use. Object—where is the heart in that? An object is something to carry around, to count, to purchase, to collect. It is something separate. The process of objectification begins with the first cut toward straightness. After the trees are felled, the conspiracy of object continues in the timber sales reports, lumberyard accounts, and architectural plans. The carpenters perhaps cradled the wood in their hands as they built this house, but did they remember the once-living trees? I wonder who among the many people who deal with wood as product have walked in the forests of these trees and listened to their voices. When the memory of tree has vanished and the connection is broken, the wood becomes corpse, or not even corpse, but something that appears to have never been alive.

The wood ceiling here is supported by two big crossbeams and pillars. I look up at these beams often because the shape and design are compelling. They form a cross. I look at this cross of wood and imagine a person suspended, connected to the wood. The image of Jesus with downcast head and pierced hands and feet evokes a powerful response of compassion. I can't help but identify with the human agony of his experience.

But what about the wood of the cross? I wonder if Jesus was embraced by the spirit of tree in his painful death. He did not die alone. Even in his last moments he was supported by life. The cross served as a connecting link to the ground, touching the common soil of our lives. Jesus on high hung partway between grounded reality and the mysterious unknown. Many have focused on the transcendent theme of his story, but what about his fundamental connection to the earth through the cross? Is this not an equally valid route to spiritual awakening?

But the cross was only a piece of a tree. Why didn't they nail him to a living tree? Perhaps that would have offered too much life force and spiritual strength. But also the tree represents intimacy; a cross speaks of exposure. This central religious story is about crucifixion of tree as much as crucifixion of person. The curving intimacy of the tree was symbolically replaced by the linear abstraction of the cross. This is the loss I feel—the living tree reduced to objectified pieces, the loss of life as it really is—vivid and unsimplifiable.

In the midst of his own tragic story, I wonder if Jesus noticed the loss of tree life. As a great teacher of compassion and a carpenter himself, I want to believe that Jesus had some care and concern for trees. It seems to me that his gospel of love applies to relationships with trees as well as people. In the story of the crucifixion, the tree did not have any choice in the decision to end its life and to be disfigured in death. The tree did not ask to be sacrificed any more than Jesus did. And the tree could not cry out to others, "My God, my God, why hast Thou forsaken me?" The recorders of history passed on the story of the great wrongdoing to Jesus, but they overlooked the story of the cross.

Gazing at the crossbeams of my house, I am caught by this double crucifixion, and with my culture I still carry the pain and woundedness of both deaths. In studying the story of Jesus, people ask themselves over and over, what does it mean? Theologians, ministers, ordinary people want to know how this suffering speaks to their lives. It encourages the practice of compassion, forgiveness, and the development of spiritual strength. The story is passed on to others, and the search for meaning stays alive.

The story of the tree is another matter. The loss has barely been noticed. The drama of tree death is repeated over and over again, day after day, decade after decade, century after century. As if in a daze, people permit the continuous execution of millions of tree martyrs, whose crosses become chopsticks, tables, paper, and buildings. I am as much a part of this web of sacrifice as anyone, and that is painful. I cannot find an easy way to live with integrity in the midst of this confusion. I am weary with wondering how much will be destroyed before we find the tree behind Jesus.

CHAPTER 14

Bones in the Land

My old street in Santa Cruz was full of potholes. Somehow the narrow one-way lane had escaped the city's watchful eye and was deteriorating gracefully. Almost invisible from the main road, the entrance to the street was obvious only to residents who knew to turn left just past the asparagus farm. The houses on the street were a mix of styles—1950s cheap and simple and 1970s spacious and rural. The neighbors on one side still kept goats, and the people next door had chickens. The ragged road with its unplanned development could hardly win a *Better Homes and Gardens* award. It had an air of being partially forgotten, partially discovered, and this confusion seemed to protect it from too much extraneous traffic.

My house was the second to last on the right. From the back door I could walk north through a hole in the fence along well-established cow trails to the western entrance of the U.C. Santa Cruz campus. From the end of the street I could walk west to the woods above the creek. I watched the sun rise from my bedroom window and saw it set from the kitchen. On the occasion of extreme color displays I would climb onto the roof for an unobstructed view of the whole grand sky.

I lived in this landscape for five years, absorbing the topography and seasonal changes through my eyes and feet. I knew the trails like well-worn grooves in my mind. The seasons revealed the cycles of eater and eaten as each species made its appearance in the local food web. First the grasses greened up in winter, filling with successional waves of soap lilies, goldfields, lupines, and poppies. Then the crickets appeared, the first ones green, the later ones brown to match the drying grasses. By peak insect time, orb-weaving spiders had prepared a thicket of booby traps on the dead grass stalks, visible only under the magic spell of thick summer fog. Garter and gopher snakes lounged about on sunny days, feasting on the abundance of spring mice. By day red-tailed hawks and by night great-horned owls hunted the fields for snakes and ground squirrels.

One trail headed downhill into a small canyon of redwoods, oaks, and bays. At the bottom, Cave Gulch Creek meandered through rocky boulder gardens, pausing in calm pools under hazel shadows. Frequently I made small pilgrimages to the creek for counseling, drawing on the wisdom of still water and rapids as needed. On warm days

I headed straight for the swimming holes downstream. Stripping off my clothes, I would plunge in over my head, shocking my system into gasps of sudden ecstasy. In winter these holes were turbulent waterfalls fraught with slippery, algae-covered rocks. But in summer the water scaled back, allowing me easy access to good basking spots. Once, in a fit of sleepless heartache, I headed for the creek at dawn to baptize my grief in cold water and send it downstream.

The open grassland stretched across the top of the coastal hills out to the bluffs above the Pacific Ocean. From the neighborhood hills the fields dropped down to the next marine terrace along the Santa Cruz coast. The base of each terrace marked a former shoreline, carved out by the waves of a once-higher ocean. As sea level fell over thousands of years, the fractured sedimentary rock was colonized by plants, creating the illusion of semi-permanent coastal hills.

Where the grassland sloped down from the ridge, the fields became oak savanna, an open dry woodland of coast live oaks. The forest was arrested at an early stage of its development by the introduction of cattle on the land. Any natural bent for complexity was then and continued to be hampered by the presence of cows. The cattle browsed on leaves, acorns, seedlings, and small oak saplings, and compacted the soil under the trees. This accounted for the absence of any young oaks on the savanna.

Most of the oaks were at least fifty years old, and many venerables had lived over a century. I was most drawn to the old ones because they offered the best possibilities for shade, backrests, and tree climbing. I had two favorites I returned to again and again. One tree was a five-minute walk from the house, quite convenient for short visits. Like many older oaks, this one had several branches dipping low to the ground. One branch actually rested on the ground; another leaned against a rock, forming a natural two-person seat. I held numerous picnics under the lovely canopy of this tree, sheltered by the intimate space of its branches.

Sometimes at night the house would seem too small for my roaming spirit and I would grab a sleeping bag and wander out to the oak tree. The first night I slept out, I discovered a perfect sleeping spot where the hollows of the earth exactly matched the contours of my body. Five or ten inches to the right or left would not do; I could always find the spot by snuggling around on the ground until I fell into place. If ever there was a personal sense of place, this was it for me. I slept out in spring under passionate moons, in summer under bright stars, and in fall under dry winds. Sometimes I fell asleep to coyotes howling and woke to a silence of fog.

My other favorite tree was over a mile away from the house, beyond the limestone bluff. Like the sleepout tree, this oak also had wide, long, drooping branches, though none touched the ground. This particular tree had made my acquaintance one day when I was looking for a quiet place to sort through the perplexities of life. Journal in hand, questions in mind, I was drawn to the solitude of this tree for stability and companionship.

The oak stood alone. The closest tree was a large broken-top Douglas fir dripping with lichens. The two together made a striking silhouette against the open hills. From up in the oak branches I could see all the way across the bay to Monterey. The infinity of the ocean was comforting, helping me place my own questions of meaning in a properly vast context. Up close I had a bird's-eye view of the wooded canyon and open grassland. Warblers and woodpeckers traveled between the two worlds while red-tailed and Cooper's hawks patrolled overhead. The tree offered many fine sitting spots, enough to host a small dinner party. Each was remarkably comfortable and secure, for the tree was old and wide, with mature and resilient branches. To sit in the tree was to sit in the arms of a solid, healthy, and well-developed organism. It was both a pleasure and a relief to be with someone so well established. From this vantage point I wrote, napped, and let go of weighty subjects. Because of the distance and the unusual capacity of this tree, I journeyed to it only on special occasions, keeping a sense of loyalty to the high caliber of our exchanges.

One spring, after five years of making my way in this landscape, it became clear that it was time to leave. I had completed my schooling, the house was up for sale, and the rent would likely double under the crush of student-housing prices. I needed to find my way in the world and move on from this womb of familiarity. The timing was unavoidable, but it tore me up to think about going. Everything I had begun had ripened; it was time to scatter the seeds. I didn't know where I was going; I just knew it was time to leave.

The sense of loss and grieving came in waves, both before and after I actually moved out. Though I knew the land like an old friend, I didn't realize how much it supported my entire life and worldview. Like roots growing up through my toes, the land had entered my body. The creek water flowed in my veins, the night stars shone in my eyes. The waxing and waning moons measured the rhythm of my blood and silence, the howling coyote cried in my deepest songs of night.

I left, finally, by suppressing the grief in favor of forward motion. This did not mean the landscape left my body or stopped speaking through me. I just stopped paying attention to what it was saying. I moved north and settled into a different landscape, investing my energy in the future rather than the past. I did not understand how thoroughly my existence depended on the evolutionary past of tree, landscape, and water. I was under the spell of my own will and determination, following the call of personal and career development.

I often returned to Santa Cruz, especially in the first year, to see friends and work on joint projects. Longing for the familiar and hungry for connection with what I knew in my bones, I would wander out to the oaks. When someone left a note on my car telling me not to park at the end of the road, I realized I was no longer a recognizable resident. My grieving for place still gnawed at me; even sleeping out under the oak did not satisfy my yearning for home. The break with place left an unhealed wound in my heart, covered over by the sequence of events and the fear of too much feeling.

After ten years I returned again, seeking more consciously the understanding of home I had abandoned. I felt I must visit the oak beyond the bluff. I set off across the dusty, sunbaked fields of fall. There was little left of the pink and purple wildflowers except their seedpods. The grasses were dry and beaten-down by cattle hooves. Narrow cow trails had become wide highways, the ruts eroded and the grass sod torn from the soil. I crossed the limestone bluff and followed the trail along the grassland above the canyon. The odd-shaped Douglas fir stood like a sentinel marking my destination. From this angle I couldn't see the magnificent oak I remembered so well. It had been a long time since I'd walked these hills; I felt slightly disoriented. But the powerful memory of the tree and my strong desire for reconnection drew me closer.

AAAaayyyyyy!!! A great wail of grief sliced through my body, the knife edge of death so present I could feel it in my fingers. The tree was gone. Someone had cut off the four major limbs just above the main crotch of the trunk. Nothing was left. Not a single branch or twig littered the ground. The tree had been dismembered, dissected, beheaded, leaving only a tall, gray corpse on the landscape. *My friend, my friend!* I cried out in shock and disbelief. How could this have happened?

My sense of returning shredded into rags of memory. I sobbed with grief, unable to grasp the stunning change before me. The tree had been stolen from the land, apparently the work of professionals. I tried desperately to construct alternative scenarios of damage by fire or earthquake, but these were clearly far-fetched. My dear friend had most certainly become firewood. I could hear the phantom conversation in the wind. "Look at this one, Joe—mighty fine tree for oakwood, don't you think? We got the truck and saw right here, might as well take her down."

On the beautiful curves of the ancient marine terraces, where the ocean filled the sky and the fields below me were fertile with autumn's harvest, they took this fine oak down. My friend, taken apart by a chain saw. *AAaaayyy!!* The instrument of torture cut through my own limbs. I looked at my hands, the hands that grabbed this trunk to climb into its canopy. I looked at my feet, the feet that walked naked on the oak's gnarled skin. I could see the ghostly shapes of the tree's limbs in the air. I could feel them in my body, in the landscape memory that had never left my cells. The tree soul in me screamed with the shock of sudden loss.

This tree left the world of the living in the hands and minds of people who did not see it as a living being with a rich history of relationship. They did not see the warblers, kinglets, and woodpeckers that nested in its branches and cavities. They did not see the ancient patches of pale-green foliose lichen on the wizened trunk. They did not see the lifetime of storms and coastal winds that blew through these branches. They did not see the underground roots interwoven with the Douglas fir ten feet away. They did not see the human friendships that grew out of the life of this tree.

This was the great sorrow—so much lost in such a short time, lifetimes of relationship dissolved by a few hours of sawing. I kept staring at the stark tombstone on the landscape, paralyzed with grief. On so many visits I have gone in search of specific trees, wanting to rekindle a friendship and take time to listen to what we knew together. It had not occurred to me that any of these trees would be dead or dying. I thought my lifetime was relatively short next to that of an oak.

In the weariness of grief I leaned up against a nearby, smaller oak. How many years would it take to grow another oak like the one that had been cut down? Perhaps one or two hundred, if new seedlings even had a chance to get started. I saw no oak saplings or even small sprouts. My longing for regeneration would not likely be fulfilled anytime soon on this grazed bluff. The sadness was numbing. I gazed blankly at the vaporized tree. Underneath the loss another deeper grieving rose up—a loss of connection with this land that was in my bones and blood, this land that I had abandoned in my search for the future.

The grieving commanded full attention. I had been in the process of becoming native when I lived here. The landscape had taught me well, cultivating a mind of place, a mind *in* place, a place of mind. When I left, I cut off this process of intermingling, of deep teaching, of learning the wisdom of time in a wild context. I had allowed myself to become dismembered. This was the greater loss; this was my complicity in the tree's death.

Sitting against the thin trunk of the remaining oak, I wondered if I could ever complete my relationship to the land. Grieving or rejoicing, how could I live this relationship to its fullest? How could I properly respect its history and complexity and carry it to the next generation? I would want to bury my bones in the land and ask that others return and renew this knowledge of place. Then some of my experience would transfer to others rather than be forgotten, and the grieving of separation would be prevented by keeping my bones close to the land.

At some level I betrayed the fallen oak by leaving. How could I expect the oak to sustain itself unharmed in a world of human use and ownership? I felt caught in the helplessness of loving trees and land owned by others with different values. I did not realize the commitment it might require to keep this friendship alive. It was, in truth, a matter of life and death.

CHAPTER 15

Overtures of Peace

QUARRELING, WHY ARE THEY ALWAYS QUARRELING? I can hardly bear the shrill voices pecking at each other. Hate barbs fly through the air like tiny daggers of destruction aimed at tender places, loaded with hurt, blame, anger. Little bombs of words go off with precision timing in all the familiar eroded places. First one explodes, then another, until the ground is littered with fragments of spent ammunition, smoking with anger.

Just another family feud, a minor scrap, an ordinary round of irritation. Comes with the territory, they say. It's human nature, don't let it get to you. How is that possible? The entire energy field in this space is fractured, unraveled. You can't walk into it without feeling the hurt and pain. A few minutes of artillery bursts and the whole thing is frayed. Tough hearts may be protected from such machine-gun attacks on the psyche. But the damage is done—I see it, I feel it. I ache with the presence of everyday violence. It is painfully familiar.

I slip out the door and leave the shattered harmony in search of stability and healing. These small acts of violence resonate with every act of violence I've witnessed—political violence, childhood violence, sexual violence, family violence. The pain stings, and I recognize again the human capacity to hurt. I welcome the relief of the sweet rain as I walk. Walk toward the trees, walk toward the redwoods, walk toward the tall ones, the people of time. The longing for their presence is palpable. I ache for the stillness and silence of their groves, free of fighting, a demilitarized zone where I can relax and be at ease.

I hop over the barbed-wire fence and walk down the trail by the creek. Branches are dripping with late spring rain. The rich, earthy fragrance of decaying leaves and moist soil is like incense blessing the air, inviting me in to the temple. Though I have not walked in these woods for several years, my feet know the way. They haven't forgotten the shape of the land, the downhill draw through the creek watercourse. I walk and walk in the rain, the blanket of mist cloaking the trees around me, washing away the sorrow and grief.

I am looking for the cave tree, a tree I once encountered by accident. Underneath this large redwood there is a small dirt chamber big enough to stand in. The entrance is easy to miss; the tree looks as solid as any other from a distance. I ache to climb into this secret hollow today and hide from the world of thoughtless violence. I want to go deep into the earth and sit in the roots of a Tall One. I am hungry for the stillness and wisdom of caves.

As it happens, the tree remains hidden. Perhaps I will find it another time. I need to stop; I'm getting quite wet. I see another Old One offering a place to listen and stand still in the turmoil. I climb into the base of the large, hollow tree stump and fit myself into its form. Inside I am minutes, hours, maybe centuries away from the quarreling and violence. The soft rain refreshes, filling the forest with gentle quiet. No one else is out walking today. I can rest in this peaceful place and let the tree energy flow into my mind and heart.

Under an overhang of bark I peer out at the rising creek. The young redwoods across the way crowd together in circles around their parents. If I move even an inch from my small, protected tree roof, the rain splatters my pages. I am standing in a roofless shelter: the old tree's top is completely gone. The hollowed-out inner walls are black with charcoal. The original, enormous redwood must have been at least ten feet in diameter. I cannot reach the other side with my hands. The forest floor has crept inside the shell, filling the bottom with soft needle litter and the wanderings of millipedes. Looking up, I see a circle of sky framed by this memory of tree, the heart of the redwood now empty and open.

How appropriate that I should be drawn to a tree whose life was cut short by human destruction. I cannot escape the presence of the mind that kills. The tree was logged for timber sometime in the last century. It had to be cut high, using logging planks above the flaring base. These large trees were fine prizes for commercial loggers; there were so many, the supply must have seemed endless. But the harvest got out of hand. Wave after wave of killing for greed took place in California—first the sea otters and elephant seals, then the whales, then the great rush to dismember the hills for gold. Finally they took the trees, racing to cut the big ones, especially those most accessible on creeks and coastal slopes. All this in half a century, leaving the ocean beaches empty, the Sierra foothills in shambles, and the redwood forests full of slashed and broken bodies.

Standing in this hollow stump, I see I have come to one who also knows something about the mind that kills. Great Hollow, I want to hear your voice. I am standing at the center of your existence, asking for your understanding of how things are. I am trying to imagine the inner life of a redwood. I want to know how you responded to flood, fire, and the invasive brutality of logging.

It has always seemed to me that redwoods are the yogis of the forest. You are all so straight and tall, with no curves anywhere in your spine, your side branches insignificant in relation to the mass and presence of your trunk. Such a powerful core of rooted energy in this tall arrow to the sun! An Old One as wide as you might easily have grown 250 feet tall and been 500 years old. I wonder if you suffered through long droughts that might have threatened the water column to your distant crown. I wonder how many fires scarred your trunk before the last cut across your great width.

These roots beneath my feet—I wonder if they grew in response to a great flood that buried your base in silt. What is it like for your shallow, widespread roots to sip the fog drip from the soil's surface? I know very little about rooting; my body parts are poorly adapted for poking through the dirt. Toes and fingers can dig in the earth, but they do not want to stay underground for long.

My back, though, knows something about straightness from meditation. Not very much, I suppose, compared with you. I am perhaps more like a willow or poplar, shifting around with every other breath. What are a few hours or days of sitting still next to a few hundred or thousand years? Unfathomably small. I practice and practice and still barely come close to a history of straightness. In this lifeform I cannot know how it feels to be straight for hundreds of years. You are a role model, then, of something I can aim for but never quite grasp.

What did it do to the quality of your mind to be straight and still for hundreds of years? I wish I could know. Something about not knowing feels helpless and sad today. There has been so much loss and destruction in the last two centuries. I carry it in my soul, this knowledge of what has come before and continues to happen now. The mind that kills is alive and well, seeing you as object and distant Other. In the pain of this I return to you for teachings. It is the returning itself that teaches. This longing to remember how we know each other pulls me inside you. I want to know you in the deep time of evolution. But even the short time frame of human history is not so easy to meet. For centuries you lived in stillness, only the occasional hoot owl or jay disrupting the silence. Through winter storms and summer fog you stood quietly in the company of other Tall Ones. I wonder sometimes if you were silent for so long that you forgot how to speak. Perhaps it was enough just to rustle with the wind and sing with the creek.

Then, not long ago, the machines came. Saws and pulleys, bulldozers and cranes. What a shock to the quiet monastery of the forest! The first timbermen were hardly prepared to deal with the overwhelming size of your trunks. No logging equipment of the time could fell, transport, or mill such huge trees. But this was only a challenge for the clever human mind, eager to sell your beautiful heartwood to trading ports around the world. In the beginning it sometimes took two axmen a week to cut through one of your relatives. Then the bark peelers would move in like scavengers and tear off your skin and branches. Fallen giants were dragged through the forest by mules and oxen across delicate violets, anemones, and sword ferns. The broken soil slid into the creeks, clogging the waterways and salmon runs with silt.

By the 1890s steam-powered pulleys reduced the time it took to haul your fallen companions to hitching points. The forest rang with the crack of axes, the whir of two-man saws, and the hiss of steam engines. The noise must have been excruciating. One by one your people were taken away, leaving only the young and twisted behind. Surely you cried out. Surely you called for mercy. But no one seems to have heard.

With the invention of the motor and the automobile, it was a short step to bulldozers, trucks, and power saws. By 1929 the rate of cutting was up to five hundred million board feet a year. The war was on, the war of the killing mind against the living. With timber trade primarily in private hands and news reports only local, it was hard to know the extent of the damage. As the degeneration of rivers and forests continued unabated, some of my people finally took notice. Efforts to stop the carnage began as early as 1852 with the first legislative proposal for a state redwood park. But it was not until fifty years later that the first protected stands were actually established at Big Basin near Santa Cruz.

The obituary report released by the National Geographic Society in 1964 revealed the shocking and enormous loss. Of the original two million acres of ancient redwood forest, only three hundred thousand remained—a mere fifteen percent of the stillness that once graced the coast of California. State parks protected fifty thousand acres as museum fragments of the once-flourishing belt of grand trees. Another forty-eight thousand acres were added in the expansion of Redwood National Park, but much of this area was damaged by erosion and logging. There is no question that the virgin groves along the coast were permanently altered by the great wave of destruction that ravaged the redwood monasteries.

I bring this sadness to you today in the weeping sky. My people have committed such great wrongs against your people. It is painful to acknowledge the scale of death in this short period of our history. I want you to know that not all humans come with machines to take you away. Did you think we were at war with you? Those who came could not hear your voices; they only wanted to take care of their own lives. You were an abundance greater than they could imagine. It did not seem possible to make a dent in the endless bounty of trees.

I cannot speak for all those who came before, or for those who still come to take you away. But as one person reaching out for connection, I ask your forgiveness. I ask you to accept this small overture of peace. I need to begin here to stop the killing mind inside. This is not easy to do. Please believe that I come in search of another way. The violence doesn't work. I am standing here in your roots asking, asking.

CHAPTER 16

Lineage of Fear

DUSK. A SPOOKY SILENCE SETTLES OVER THE FOREST. I am here alone, or at least I think I'm alone. The creeping darkness slips in through the trees, wrapping its fingers around each craggy bough and fallen branch. Evening shadows thicken, stealing the space between trees, taking over the forest, taking over the path I walk on, erasing the distant perspective.

When I entered the forest earlier, the late-afternoon sun still penetrated this dense stand of Douglas firs. It had taken half a day's drive to find this small fragment of ancient forest. I had only a few sentences in a guidebook to direct me to this new territory. Despite years of childhood exposure to Douglas firs, I had never seen a stand of old trees. My Oregon tree education was limited to suburban plantings, Christmas trees, roadside scenery, and forests in a few state parks. No one ever mentioned the treasure heritage of great trees—not in biology class, not in church, not in scouts, not in camp, not at home. Nobody told me about this central feature and evolutionary gift of the Northwest landscape, the great conifer belt on the west side of the Cascade Mountains. No one introduced me to a single Douglas fir elder in my childhood.

So now I am catching up, trying to make up for lost time and meet the Douglas fir forest properly before any more of it disappears. In this short snatch of time, I am asking this two-acre piece of untouched history to tutor me. Other fragments are much less accessible, requiring long drives on winding logging roads through clear-cut country. All the lowland forests have long since been cut and milled for timber. I consider myself lucky to have this small chance to learn about my native bioregion.

The light of late summer does not linger. Once the sun drops out of sight, the forest edges begin to soften. The uneven canopy casts ragged shadows over the forest floor. I follow the trail to a single old fir. Its swirling, oversized branches dominate the space, reducing the possibility of any other trees growing to large size underneath. Delicate vine maples and hazel dance in the light that slips through to the understory. Thin stems of vanilla leaf and inside-out flower poke through the brush.

In the dimming light the tree is a bit spooky. Compared with redwoods and other tall conifers, old-growth firs do not have clean lines and perfect symmetrical forms. The

trees drip with hanging mosses that seem to rot their branches out from under them. Even before they hit the ground, the decomposed limbs have lost their twigs and needles and are well softened by insects and fungus. The broken edges of fallen branches give the forest floor an untidy appearance, as if the trees were self-destructing. The satisfaction of a crisp, dry, orderly ponderosa pine grove does not exist in this moist temperate forest. The place is messy.

Unkempt as it is, this great tree structures the shape of the forest here. One might call it the dominant tree, as it takes up the most light, space, and water. But dominance implies rule, and I am not sure that a tree can rule over a forest. That the tree has the greatest presence in this area is clear. And in its role it naturally influences others around it, like all beings of great presence. One might call it the Great Stabilizer or the Great Fertilizer or even the Great Giver. Instead of posing the tree as the Great Consumer of resources, one could see it as the Great Provider. Fog drip, litterfall, rotten branches, underground soil stability—all are gifts of an old tree to a forest. The tree shapes the history of the forest quite literally as the cells of its own body shift from tree to soil, canopy to forest floor.

I walk farther into the forest, watching the color of life fade into darkness. I try to read the tree silhouettes as I follow the trail, but it is difficult to make out details with any precision. My eyes strain for light, growing tense with the uncertainty of being in a strange place. I stop to listen for other footfalls; my heart is suddenly very loud. I am no longer certain I'm alone. What if there is someone else here? I am so vulnerable, so unprotected. I can hear nothing but the creak of branches and the rustle of leaves. The emptiness is eerie.

I walk on slowly, quietly, placing each step with care and lightness. The night world has taken over. Under the command of evening I become invisible, inaudible, holding my breath and disappearing into the shadows of the trees. All my information-gathering faculties are focused on the forest. Shapes, sounds, movement, texture of ground underfoot—I need to know it all; I am on alert for the unexpected, the dangerous, the life-threatening.

In the shapeless hours of twilight, a small, nagging fear takes hold. Fear of meeting a strange man in a strange place in the spookiness of dusk. Fear of being raped. This is a fear born out of experience, born out of fleeing from strangers twice in forests, far from safety. In both cases the demand for sexual contact was unmistakable. The small of my back tightens; I stop breathing. Listening, listening—I need to know, is anyone here?

The fear, now arisen, has grabbed my imagination. I no longer have control. I know this fear, I recognize it from many lifetimes as a vulnerable animal. The fight-or-flight response of adrenaline is built into my system to protect me from danger, to urge me to respond, to ensure my survival. In a single turn of mind, the strangeness of the forest and the unknown has become threatening. I don't know the sounds of this forest at dusk; I don't know who inhabits this land, who comes here to visit, to hunt, to eat. I don't know the life of this forest in this season and time. I am a stranger.

All the fears come up together. Fear of getting lost, fear of enemies, fear of attack—I know the terror of being pursued by threatening men. Fear of the dark, fear of death—I know the engulfing shape of the unknown, the archetypal myths of annihilation. Fear of the collective unconscious, fear of the shadow—one dragon after another parades before me, stealing my breath and stiffening my stride. The fear encoded in my cells and nervous system makes a pit in my stomach.

I can no longer see the beautiful forest. I only respond to what I can't see. With adrenaline pulsing in my arteries, my whole system stands ready to protect me from all that might harm me. Each crackle and snap rivets my attention. I cannot remember even the little I know about this forest. The long history of biological fear speaks its own truth through my frail and uncertain body.

Yes, in this fear I would want to cut this forest down. Yes, in this fear I would feel safer with no craggy, broken, messy, murky trees blocking the light. Yes, I would feel safer if I could be sure there were no strangers lurking behind these wide, cumbersome, unpredictable trees. Yes, I know I'm afraid. Yes, I am a stranger. Yes, right now in this moment I am the odd one, the one who doesn't belong, the one whose life is at risk. In this moment of adrenaline surge I will do anything to protect my life. I am both paralyzed and galvanized.

This much energy is hard to contain. It begs for release, for expression, for the chance to do something—anything! I can barely restrain myself in the silence. Fear is my ruler; it issues instinctual commands. I am at the mercy of this ancient fear, this wild and powerful fear of being consumed. Consumed by strangers, consumed by animals, consumed by the Other. And this fear right now is the fault of the forest. I would not be afraid if it were not for the cloying darkness gripping the tree branches, calling out the night spirits, the spirits of the dead.

Dead branches, dead millipedes, dead crickets, dead wood rats caught in the talons of night owls. The death of the forest comes alive at night, speaking in the snags and rotten logs where bark beetles burrow through the decaying wood. Death is everywhere in this forest, I can taste it. It raises the hackles on my back; it tightens my scalp, it tenses my fingers. Do I want to meet this forest on its own terms? Am I willing to make company with death on *its* own terms? No, I am not certain I do. I just want to feel safe.

The consuming presence of fear erases relationship, erases rationality, erases the possibility for knowing this forest as it stands. The powerful message of adrenaline is *LIFE! LIFE! Stay alive, stay alive!* The original cause for fear, grounded or ungrounded, is now lost in unintelligible paralysis. My only thought is, "Get out, get out of this place before something happens to you."

With a strange twist of gratitude I reach the fire road, now safe from the darkness of the forest. Where all the trees have been cut down, I can see the sky. I can feel the starlight, I can walk without stumbling. It is a great relief. I run for the sheer release of tension, for the playing out of the natural evolutionary response. I run to get away. I run to escape the taste of death. I run to breathe again. I run to make peace with the fear.

When the European settlers came to this continent over three centuries ago, they encountered a forest they didn't know. The trees were tall and mature, like none they had seen in Europe for generations. Wilderness had long ago disappeared under the sweat of wars and the privilege of class. The European landscape was ruled by the metaphor of garden, of orderly fields and tamed landscapes. Those who arrived on the western shores of the Atlantic brought this metaphor with them and planted it confidently in the foreign soil.

The forest was an impediment to the garden. And it was strange. The native people who lived there were strange. The settlers' survival depended on their response to the Other—to trees, birds, mammals, and people who were different. A large part of that response was fear. And a large part of the fear was covered over by desire—for land, for wealth, for success, for adventure.

Eventually desire triumphed over fear. The cutting of forests was equated with security, prosperity, and the advancement of civilization in the New World. Social and religious values sanctioned the acceleration of this destruction. Desire for more, more, more drove the fear underground, paving it over with cities, farms, and factories.

But each new wave of expansion brought a new wave of fear. Every forest was home to strangers, from the point of view of the settler. Many of these strangers were willing to fight to protect their living connection with the land. And the fighting brought more fear. Fear of torture, fear of kidnapping, fear of murder. For many settlers and the government, this fear was overcome by fighting back, sometimes with massive displays of cavalry and gun power.

When the European strangers reached the Far West, they saw an apparently endless forest of magnificent spruce, cedar, redwood, hemlock, and Douglas fir. The original conifer belt extended two thousand miles from the Alaska panhandle to California's Golden Gate. Many of the trees were fifty feet around and over three hundred feet tall; some were more than two thousand years old. The forest of giants covered seventy thousand square miles along the backbone of the Pacific crest down to the ocean. And in the Pacific climate of wet, freezing winters and hot, dry summers, the conifers thrived, outpacing the vulnerable hardwoods by photosynthesizing in winter and minimizing water loss in summer.

In response to the overwhelming abundance of trees, desire went crazy. Piece by piece the forest was taken. Fear was banished in a gaudy celebration of stolen wealth. In the race to strike it rich, there was no room for cowardice and hesitation. The strange forest was tamed by cutting it down and turning it into money.

I know this history from walking through pieces of forest in the Cascade and Sierra Nevada ranges. I know the stories of private industry and government agencies that accelerated the cutting of the forest. But what I want to know is the story of the fear. In the three hundred years that settlers have been cutting trees on this continent, it is not clear that there has been any progress in learning about fear. It may even be true that fear has accelerated, concentrating the destruction of nature on fewer and fewer acres of forests.

In the tiny fragment of old growth forest I visited, I felt that concentration of fear. It was more than personal; it was historical. It was full of information. It was mature with the history of death. I am certain that I cannot fully understand the loss of ancient forests if I do not understand this fear, because all of this history is my history. I am a person living on this continent as a North American; native or not, this is where I was born. I am part of the lineage of settlers who came to a strange place and experienced the fear of strangeness and the threatening Other. I need to know how this fear drives our actions unconsciously. I need to acknowledge the powerful role of fear in conquering the craggy darkness of the old-growth trees.

The fear can no longer stay underground. It has sprung to the surface through a leak in the system. And now it takes a different form—fear not just of personal survival but of planetary survival. Fear of losing the whole thing. The dark shadow of fear is roaming the broken landscape, loose and unattended. Loss of one regal Douglas fir of ancient character is a small loss compared with loss of the whole forest. Loss of one's own life is a small loss next to loss of the whole tribe. Now the fear demands a response. The smell of death is palpable in the shaven ridges and clogged rivers.

PART FOUR

Finding a Way

CHAPTER 17

Pilgrimage

WHEN THE BUDDHA ARRIVED AT BODH GAYA, HE had been on a long journey in search of the truth. Wise men, hermits, and ascetics had offered him their teachings, but he was not satisfied. Though he had disciplined his body to the point of starvation and challenged his mind to the edge of its limits, he still had not found the deepest truth of existence. His determination burned in him like a steady flame; with all his heart he wanted to penetrate the way to true understanding.

In his passion and yearning, he was called by a large fig tree to sit at its feet. This was the destination of his pilgrimage; he had come with his empty body and mind to receive great wisdom. He vowed to not move from the tree until he found the Great Truth. For seven days and seven nights he sat in meditation, supported by the tree. During the seventh night he was tormented by every possible distraction of the mind. Mara, the voice of delusion, challenged him ferociously, asking what right he had to sit by the fig tree seeking the truth. To counter the force of ignorance, he touched his right hand to the earth. With the earth and tree as witnesses, the powerful realization of interdependence was revealed.

The Buddha's awakening was born out of his tremendous effort and preparation. One undertakes a pilgrimage such as his to seek the deeper meaning behind the ordinary. Because the details of everyday life can be all encompassing, one must make an unusual effort to go beyond the familiar. A journey for truth prepares the pilgrim to receive the gifts of insight and revelation. The greater the distance and time one invests in pilgrimage, the greater one's desire for truth.

Of all the trees in North America, the giant sequoias receive the most pilgrims each year. People come to the famous groves at Yosemite, Calaveras, and Kings Canyon from all over the world; everyone has heard of the Big Trees, the largest living beings. They come as tourists, botanists, environmental devotees, curiosity seekers—all of them travelers making a special effort to see these trees. The magnitude of the huge redwoods far exceeds the scale of all other trees. By simply arriving, a pilgrim enters the realm of the extraordinary.

I am one of those pilgrims, awestruck in the North Grove at Calaveras State Park. I stare at the mammoth trees, guidebook in hand, unable to grasp the words I read. The largest tree in the grove, known as the Empire State, is nearly one hundred feet around at the base. One hundred feet! It would take twenty people, arms outstretched, to make a circle around this tree. I cannot see the top of the crown; even the lowest branch is ten stories above my head. This enormous pillar may weigh twenty-six hundred tons, the equivalent of a small ocean-going freighter or eighteen great blue whales. I can hardly believe what I see.

When pilgrims go on journeys, they often return with tales of wonder and astonishment. Something of this sort happened in 1860 when the Calaveras Big Trees were discovered by A. T. Dowd. He was hunting grizzly bears at the time. When he returned to camp, no one believed his tale of the trees. Awe is not a transferable emotion; it must strike each person individually as he or she is moved. But Dowd convinced his comrades to make the twenty-mile trip through the mountains so that he could show them what he had found. Dowd's desire to share his wonder lies at the heart of all pilgrimage traditions.

Strange things started to happen, however, when redwood speculators tried to transport the experience of awe to the masses. Downed logs were apparently not big enough to convince the skeptics, so the showmen selected several of the largest trees for sacrifice and export. It took five men twenty-two days to fell a single tree, using pump augers and wedges. The foot-thick bark was stripped from the tree in chunks and then reassembled at exhibition sites, where the curious paid money to be impressed. Environmentalist John Muir was angered at the preposterousness of skinning a tree to reveal its greatness. While the empty shells of the trees went on tour, the awe was left behind in the grove. The travesty was further compounded by planing one of the remaining stumps for a dance floor; the fallen trunk became a platform for a two-lane bowling alley and saloon.

All of this history is part of the interpretive tour through the North Grove. Though often I despair over human shortcomings with trees, today I am a pilgrim with an open heart. Human influence is only a small part of the three-thousand-year history of these sequoias. My journey is a search for the depths of truth held by these magnificent trees. I want to hear as much of their story as I can.

My guidebook provides information about the features of a giant redwood—bark, needles, cones, and life cycle. It suggests a pilgrim's route with twenty-five numbered stops, twenty-five chances to meet the Awesome and be moved. At the first stop there is a platform around the base of two trees growing side by side. It is big enough to hold a crowd of thirty people. Climbing the steps of the platform, I enter the temple of the trees like thousands before me. A fire or perhaps a number of fires have hollowed out a cavern between the two giants. I lean into the opening, rubbing up against the soft spot on the bark created by countless communions between tree and human. The fibrous bark is almost polished as if a waterfall had been streaming down the tree.

Across the hollow I can see the inside of the seam where the two trees have pushed up against each other. The bark has merged, obscuring the original separation of their bases. Like a double star, the energy fields of the trees are joined, swirling around them in overlapping circles. I try to imagine the contortion of the wood at the base, compressed from the strain of carrying over a thousand tons of biomass. The sheer physics of this structural challenge require the tree to aim for perfect proportion and weight balance within the base, trunk, branches, and crown.

The experience of the Awesome is unnerving. I feel myself shrinking to insignificance next to these giants of time. My less than half-century lifetime is nothing compared with their several millennia, my five-and-a-half-foot frame a mere two percent of their height. The giant sequoia scale of reality is so big next to anything I've ever encountered that my mind simply equates it with infinity. Too big to comprehend, too big to grasp with any ordinary mental process. Yet this is the task of the pilgrim—to go beyond the limits, to jump off boldly into the unknown and then return to the known world, enlivened by the encounter with the Awesome.

In many religious traditions the way to spiritual growth and development is seen as a path, traveled by many people over time. The spiritual path is one route to awakening from the sleep of unconsciousness that plagues human beings. A pilgrimage is a more intensified part of the journey, a time to be tested by the unexpected, to ask questions of meaning, and to gain strength and merit. Islamic pilgrims walk the long route to Mecca, Catholic pilgrims go to Rome. Jews travel to Jerusalem, Buddhists to Bodh Gaya, India. All these destinations are now urban cultural centers. In contrast, native peoples make pilgrimages to sacred places on the land, reconsecrating them generation after generation by personal visits and ceremonies.

Fortunately, the giant redwoods have not been claimed by any single spiritual tradition as a sacred site. The temples are accessible to visitors from all religious and nonreligious backgrounds; anyone may come and experience awe and humility in these groves. Most of the well-known sites are held in the public trust, protecting the opportunity to pay respect to living beings older and far more enduring than oneself. Offering homage to the trees, one cultivates ecological virtue, gaining some sense of the evolutionary miracle that spawned human life.

The North Grove at Calaveras is one of seventy-five temples in a 250-mile section of the Sierra Nevada. The trees grow primarily in groves rather than as isolated individuals, preferring moist sites at five to seven thousand feet on the western slope. The relative rarity of the groves, existing as relict populations from an earlier, wider distribution, adds to the sense of encounter with the unusual.

Today's sequoias are limited at higher elevations by harsh winters and at lower elevations by lack of water.

One could compare these groves to the ancient temples of Athens, or Machu Picchu, or Stonehenge. But the redwood trees are not ruins, they are still alive. Each grove is part of an unbroken gene flow originating millions of years ago. Each mature tree is a living history over two thousand years old. These are very old churches. But the sequoias do not look old or ruined; they radiate a dynamic vitality that dwarfs all other trees in the area.

A visitor to these sacred groves might be lured into believing that sequoias are always still and undisturbed. But there is a secret here. The true nature of these trees is intimately related to fire, element of spirit. The fibrous, thick bark wraps the tree in a fireproof robe; yet the seeds depend on fire to germinate. Without fire to clear the ground, the trees do not regenerate. The seeds are deceptively small, over ninety thousand to a pound; large trees may hold as many as forty thousand cones. In the life of a sequoia, decades might go by with no new seedlings. But then, suddenly, fire races through the grove, leaping and burning. The trees dance in the heat on the edge of death and rebirth. High above the roaring flames the ground litter crackles with the snap of needles and branches. The fire takes the tree to its edge, the fire is in the tree, the fire and tree are one thing. The temple burns with the ancient flame of spirit—bright, majestic, and awesome—and the seeds prepare to sprout.

A pilgrimage to the Big Trees is a chance to remember the capacity to burn with the fire of existence. Seeing this, one can recognize the same capacity within oneself for new life reborn of passion, for clearing that brings renewal. The blessing of the pilgrim is the journey itself. To align one's journey with the power of trees is to reach beyond the ordinary to the temples of truth.

CHAPTER 18

The Attentive Heart

BREATHING IN, BREATHING OUT. SLOW DEEP INHALE, slow deep exhale. Quieting the body, quieting the mind. I woke up this morning under the graceful, arching branches of bay laurels and Douglas firs. All night the trees have been conversing under the full moon, weaving me into their stories, capturing my dreams with their leaning limbs and generous trunks. Breathing together as I slept, as they rested, we danced quietly in the summer night. Their great confidence framed a circle for my waking; their sturdy presence offered an invitation to be still.

I arrived last night to join others on retreat in a small community in Anderson Valley near Mendocino. On this flat, gentle river bottomland the trees have grown up in easy conviviality, nurtured by floodplain water and the protection of the valley. Below the knoll the creek winds its way through a lazy channel, limpid with the slow movement of late summer. The central grassy area is open and spacious, framed by the comfort and stability of trees. Tall, straight redwoods and firs emerge above the rounded coast live oaks, bays, and madrones, filling the sky with quiet companions.

Inside this large ring of trees lies an island of stillness, a protected area in a war zone. These several hundred acres have been designated for slowing down, for listening to the calls of the heart. Their purchase was an act of intention on behalf of trees and people, that they might find a more peaceful way together. Up and down the valley, stands of redwoods are being turned into lumber and cash at an alarming rate. The tension over trees in this county is palpable. The high price for rare clear-grained heartwood is a driving force behind more and more logging. The economic machine justifies and perpetuates the killing in this war. For some the price is too high, since logging also causes fragmentation of wildlife habitat, severe soil erosion, and widespread loss of salmon runs. The battle involves private property rights and the defense of ecological integrity. Tree lovers prefer the trees alive; the timber companies want them dead. The two desires are completely incompatible.

The meditation retreat, however, is not about trees; it is about the attentive heart, the heart that feels the presence of others and the call to respond, the heart that lives in relationship with other beings. The attentive heart is not a purchasable item; its value cannot be measured in economic terms. The capacity for compassion and response grows slowly from cultivation and practice. In this retreat we are practicing the traditional Buddhist methods of mindfulness and intention. Breathing in, breathing out, with awareness, over and over again, we are trying to pay attention to what we are actually doing moment to moment. The instructions are simple, but the practice is very difficult. The mind is so naturally slippery, so deftly agile, so quick and ready to dart off in any new direction. To notice even ten breaths in a row seems an impossible task. Like practicing scales on an instrument, watching the breath can be tedious, even boring; in this lies the great challenge to keep coming back, to keep trying to settle the scattered mind.

Though there is no escaping the local tree war, I find it stabilizing to focus on one activity, one motion at the center. Breathing slowly, the monkey mind finds a place to rest, to empty out, to pry loose from the paralyzing traps of self-absorption. After an hour of sitting silently, we step outside for a period of walking meditation. Each time the mindfulness bell rings, we pause and breathe deeply three times, noticing the detail of where we are. We walk so slowly, it actually makes me laugh. The retreatants look odd drifting across the lawn like misplaced jellyfish or banana slugs. One step, breathing in, one step, breathing out. Paying attention to the feet, paying attention to the breath, noticing the body moving through the landscape.

I walk with bare feet, soaking up the sunlight in the grass, crinkling the green leaves with my toes. In the center

of the soft lawn I bump into the roots of an old Douglas fir stump. A tiny oak seedling has taken shelter in a crack of the stump, drawing on the tree's remaining nourishment. The tree roots protrude a few inches above the ground, marking the space of its former water territory. Worn and smooth, they are like firm hands touching my feet. My feet, the tree's feet—we meet each other in the deep breathing that connects body to ground. I touch the tree's presence by walking the length of its roots. Next to the ephemeral exuberance of the grass, the roots provide depth and grounding, a testimony to the history of the tree.

In the slow time of meditation I practice observing each sound with attention. A bumblebee on the lawn works the tiny plantain blossoms, methodically gathering the morning pollen with self-absorbed buzzing. A large blue dragonfly whirs through the open air. Up in the trees scrub jays squabble with squirrels over territorial rights. Acorn woodpeckers call back and forth, scolding intruders. Each sound is surrounded by a generous spaciousness. Each sound is connected to a specific individual and event. In the silence of walking I hear each relation.

Cultivating this practice of mindfulness is painstaking and demanding. In each moment of observing a leaf, a squawk, a firm touch, there is the temptation to make it something more than it is—an object of fascination, a delirium of nature bonding, a symphony of deliberate orchestration. There is also the danger of thinking it something less than it is, missing the context and history of the tiny event striking the senses. Either way one falls off the impossibly thin razor's edge of bare attention. Fall and return, err and correct. Like riding a bicycle, the mind aims for balance, seeking to stabilize the wobble between the pulls toward falling.

Each step, listen, breathe. Each step, note what is actually happening. It is difficult to hold the tension of these instructions in my body. I want to run away from them, hurl myself horizontally through space rather than drop vertically through time. Slow people moving like molasses on the lawn—we are all so serious about this! Couldn't I go up and tickle someone? Wouldn't it be fun to break their attention with peals of laughter? I feel impatient and mischievous with the slowness of this practice. Breathe, relax, observe the mind of resistance. Slowing down again, I walk with grass, roots, sky, clouds, watching the emotional waves rise and fall, surge and pass away. Emptying out of self-referential ideas, emptying out of the tendency for distraction, I am trying to maximize the possibility of being completely here. But every second there is a tug in the web that pulls on my attention.

Loud, heavy, gear-grinding, gas-guzzling noises invade the island of stillness. My body tenses. I recognize the sound of a logging truck on the local transport route between forests and cities. I know more than I want to about the sound of this logging truck. The roaring engine sets off an internal alarm tied to fear, protectiveness, uncertainty, helplessness. *The forests! The forests!* the voice of concern calls out. Breathe, walk, listen, observe. The tension sinks into my stomach and tightening hands. I try to stay present to the whole causal net, to the desire to escape it, to the tension of the conflict, to the sense of threat to my survival. I know the trucks are carrying trees stripped naked into logs, their arms hacked off and left to rot or burn. I know that a logging operation can quickly turn a living forest community into an unofficial burial ground. I imagine the trucks as hearses in a long and very drawn-out funeral procession. A wave of great grieving washes over me. I struggle with this slow walking, torn between acting and not acting. It seems like an indulgence to take the time to cultivate mindfulness when so much is being lost.

But this is the tension—to find a considered way of acting not based on reaction. Building a different kind of sanity requires a stable base for careful action. It means being willing to know all the dimensions of the reality of destruction, being willing to breathe with the tension of emotional response, being willing to cultivate tolerance for unresolved conflict. This nonverbal form of ethical deliberation depends on the careful work of paying attention to the whole thing. Meditating, walking slowly, calming the mind by centering on the breath—these painstaking, deliberate practices increase the odds for acting intelligently in the midst of crisis.

The bell sounds to close the period of walking meditation and to begin the break. I am longing to shake off the tension of the logging dilemma. Between the orchard and the kitchen, a small path drops over the hill and winds through a sloping oak woodland. I follow it intuitively toward the low places, hoping to find water and the company of alders. To my delight my first encounter on the floodplain is an exquisite and abundant community garden, source of our soups and salads. Six-foot tomato plants droop with the weight of juicy red fruit, a fencerow of peas hang ripe for the picking. The quiet eye leaps for joy at the brilliant orange and yellow chrysanthemums, coreopsis, and poppies. After a morning of silence and restraint, my senses feast on the stimulating sight.

Past the lettuce, past the eggplant, past the zucchini, I aim for the faint sound of water over rocks. My feet want to stand in cool water, my hands yearn to splash wetness on my face. Stepping over the cowpies and fallen oak twigs, I leave the path and wander down to a shallow stream. Warm and almost stagnant, the water is barely moving. Near an overhanging alder the creek is a foot deep; I slip out of my meditation clothes and into my fish body. Wriggling, squirming, splashing, cleansing—for a few moments the existence of suffering is a distant thought. The tension of human confusion slides away; I bask in the apparent simplicity of animal life.

Stretching out in the midday sun, I let go of the strain of knowing so much and paying attention with such discipline. I snooze on the warm rocks, resting like a lizard. Wavering on the edge of consciousness, my mind drifts with the sounds of the stream and the warmth of the sun. Thoughts skim across the surface, finding no anchoring place in the pond of my imagination. The tension of acting/not acting is swallowed up in a yawn as I turn on my back to face the full sun.

By late afternoon we have been sitting and walking silently for several hours. I fight it less, willing now to just do the practice, just put in the time. The logging trucks still roll by with disturbing regularity, but the day has ripened, slowing my reactivity and emotional responses. The accumulation of warmth and sunshine has softened the field of green bordering the trees. My companions walking slowly across the lawn seem more like trees than people; they are less awkward, more comfortable, less ruffled around the edges. We are absorbed in the practice of remembering where we are, remembering our relations, noting the suffering of ethical tension. It takes time to see the deeply encoded patterns of destruction and transgression against trees and other nonhuman beings. It takes time to cultivate a relational sensitivity that is compassionate and not pathological. It takes time to embrace wholeheartedly the complexity of living with trees.

I find some comfort in our communal clumsiness. We each stumble along the uncharted path. Practicing with others is a useful antidote to the isolation of insight. We walk together sharing the silence, giving each other support as we investigate our lives. We forget and remember, moment after moment, each of us making an effort to deepen our capacities for observation of self and other. By learning in community, we practice breathing in a circle of friends and companions. Against the backdrop of ecological uncertainty, this retreat seems like a very small contribution of attention. Though I cannot know how it will affect the large-scale patterns of social relationships with trees, I make an effort anyway. The choice to practice awareness, over and over in each moment, is the cultivation of intention, a quiet, fierce kind of passion that supports the capacity to act with restraint.

Old tree stump, young oak sprouting, jay, and woodpecker—with your company I am just breathing, just walking, trying not to stumble on the irregular terrain. In this steady silence I ask for help to walk more gracefully, for patience to cultivate an attentive heart.

CHAPTER 19

Cutting Wood

WHAT IS MY RELATIONSHIP WITH WOOD? I CARRY this question like a burning coal to the jumbled pile of firewood that needs stacking. It is only a slight variation on the question I have been carrying for years—what is my relationship with trees? The questions serve as Zen koans—teaching puzzles not meant to be explained by the intellect but used instead to penetrate the nature of reality. Koans work on the questioner like a mantra or meditation, unsettling the mind to open up new ways of seeing. Each piece of wood presents this koan in material form; the jagged heap challenges me to pay attention to the question.

A mixed cord of red cedar, coast live oak, and almond has been delivered to my doorstep. I look at the vibrant green moss on the oak and wonder where the wood came from. After five months without rain the moss should be pale and dry. It's hard to believe this oak has been sitting in a woodpile curing. I suspect it was imported from moist forests in northern California or perhaps from as far away as Oregon or Washington. This means that local firewood harvesters are going farther and farther away to cut wood for Bay Area stoves and fireplaces.

The cedar in the pile raises similar questions. Some of the chunks are sixteen inches or more from core to bark. If that is the radius of the trunk, the source tree must have been almost a yard in diameter. I know of no trees that size in the San Francisco Bay Area. Again I'm concerned. Is this wood coming from the Sierra Nevada, North Coast forests, or even Oregon?

Both observations are disturbing. It means that the local supply of wood is probably being overharvested. Or that people are cutting wood farther away, where wood is cheaper, so they can make a greater profit. These thoughts bounce back at me, naming my ignorance, revealing my helplessness as a consumer. I only ordered the wood; I didn't collect it myself. I am just one small piece of the market equation that determines the fate of firewood trees.

Whatever their origin, these pieces of forest are now building blocks for my woodpile. I ask them more questions as I frame the foundation row for the stack. The koan deepens; it works its way into the details of each shape and form. Chunk of oak, where have you come from? Soft red cedar, what hands cut you down? Twisted almond, who

walked through your orchards? I talk with the wood, using the questions to begin a relationship that will serve me through the winter. The pieces have been cut with precision. The edges are clean, not ragged, and most of the chunks are the same size. This makes them very easy to stack. I offer gratitude to the cutter for paying attention as he worked.

Wood stacking is a labor of love. A woodpile is an art form. People who appreciate this recognize each other by the shapes of their woodpiles. A fine woodpile, like a good stone wall, reflects the eye of the stacker. Each piece is handled with a particular feeling for its placement. Where exactly does it fit in the developing sculpture? I consider each piece in my hands, looking over the stack for just the right spot. If it lies snug and stable, then it will be secure enough to support others on top. I touch and notice each piece for what it is, another form of tree. We have a silent communion as I place piece after piece on the pile.

In the silent rhythm of work I listen again to the koan—what is my relationship with wood? The questions fill my mind: What does it mean to consume wood? What is my responsibility to trees as a consumer of wood? What is the impact of firewood harvesting on California oak forests? My actions are part of a complex web that I can ignore or pay attention to. Paying attention is far more difficult and demanding. The questions obligate me to engage at a deeper level. I want to work with this wood, not against it.

I consider the internal structure of the pile. How can I stack this wood with an eye toward harmonious relationship? I see that with these pieces of forest I am building another ecosystem. Already crickets and lizards are seeking out cracks in the pile. Pillbugs and slender salamanders will crawl under the bottom logs and hide there over winter. Perhaps a mouse or two will settle on a wider shelf between pieces. Leaving cracks and holes for animals, I lay the wood down, thinking of those who will inhabit this woodpile.

A woodpile is a system on a very small scale. This makes it more than just a sculpture: it is a living art form. In fact this is what all ecosystems are—living art forms. As living forms of art, these systems are more complex and multidimensional than a human mind can imagine. That is the beauty of it—the system of comings and goings that make up this architectural event in time. By stacking the wood with some eye to system, I touch a little of the bigger story.

Still the koan is not fully answered. I carry it to another cord of wood a week later, where I am plunged more deeply into the question. This wood is bay laurel from trees thinned for fire protection. Now I am the one cutting the wood; it is my hands on the chain saw. It is my fingers on the trigger making the whirring and biting sound. Can I stay conscious and aware of what I am doing? What conversation do I have with these trees as they fragment into firewood in quick, efficient slices? The questions penetrate my body with the raw force of the saw.

I cut the limbs with a requesting heart. It is a cold day, and I need the warmth from these trees. I am asking them to serve my life. I am asking them to enter my bones and blood and fuel my cells with fire, to make it possible for me to stay warm through the winter. They say that firewood warms you twice—once while you are chopping it and again while you are burning it. But I think the real warmth comes from the heart's genuine request. It is a request for relationship—for direct, intimate, interpenetrating relationship. In this case it is the relationship of one organism consuming another, one life sustaining another.

As the chain saw vibrates, the resistance of the bay branch enters my right arm. It shakes and bumps as the tree and I make contact. Again I wrestle with the questions of relationship—tree or wood, will you kindly serve me? Will you accept my gratitude for your life given on behalf of mine? May I know you through the exchange of energy and warmth? The questions are a bridge of connection even in the process of fragmentation. As the wood falls off the sawhorse, I imagine flames consuming the chunks. Each piece burns fiercely, purifying the questions, stripping bare the questioner.

The hungry chainsaw sputters. I stop to refill the tank with a mix of gas and oil. Gas and oil from where? The Persian Gulf, Alaska, or the coast of southern California? Through what war zones or ice floes has this gas traveled before entering the chainsaw? How much has it already cost in transportation and energy to produce this liquid gold? The questions multiply as I sink into the core of relationship. The koan digs into conditioned thinking, unconscious patterns, habitual ways of seeing. It works me like a teacher, opening the possibility for insight.

Each cut requires a diamond mind—sharp focus and attention on the wood and saw. This is dangerous activity; one false move could land the saw in my leg or forehead. Mindfulness is not something to dabble in here; it is sheer necessity. I ask each piece—how will you respond to this saw? Where are your knots and hard places? How can I be most attentive to your shape and form? This is the artist's question—how does one work with the materials to honor them? How do I become the material that is being worked?

The subject shifts. I am no longer just listening to the wood. I am engaged completely in this relationship; I am meeting the tree with total presence. The chainsaw brings us to the point of intimacy, the hinge point around which all aspects of the story turn—fire, woodpile, oil, mind, danger, connection—each interpenetrating in the meeting place of our bodies.

The brilliance is tiring, the physical meeting of tree and person is so magnified by the chain saw. It requires such tension to grip the saw and manage its behavior. My hand is shaky, my arm aglow with electrical energy. In an instant of falling away, my mind slips off in a thought, and I can see the possibility of accident. Away from the riveting meeting ground of intimacy, the whole thing falls apart. And it happens in a single, drifting thought. Now I have become part of the woodpile. The tree is in my body; we have met through the passionate medium of the chain saw. The art form of wood stacking rests on this knowledge of woodcutting. The koan pierces through all the elements, burning a flame of naked insight in the core of our meeting.

CHAPTER 20

Held by a Living Being

SUN ON MY SHOULDERS AS I WALK INTO THE LATE afternoon, soaking up the sweetness of summer, basking in the spaciousness of the open landscape. Walking, I have been walking now for miles across the grassy hills of the inner coast range, just walking. Letting the landscape fall into the rhythm of my steps, letting the sun penetrate to my inner core, just walking. Letting the mind rest from the difficult work of embracing the fullness of human-tree relationships. Just resting as I walk.

From the ridge I look across Knights Valley to Mount Saint Helena and Table Rock, setting myself in place by the landmarks. To the west lie the Santa Rosa Valley and the hills stretching out to Bodega Bay. Golden grassland and dark-green forests follow the curves of the land. My eyes gaze on the long view, comfortable with the absence of obstacles, relaxed in the familiarity of infinity. Along the rounded silhouette of the soft earth body, the afternoon fog appears like a silk scarf gift for a holy lama. The day hovers at a pause point in the stillness of the afternoon warmth. Walking, I keep walking across the landscape, drinking in the wealth of distance and perspective.

On this walk my feet are leading the way, following the rhythm of rest in motion. Settling into the economy of streamlined movement, I relax into my hips and thighs, trusting the intelligence of contact with the ground. My animal feet press lightly on the earth, prancing over tree roots, alert for swales and hollows, darting around rocks and stones. With the implicit understanding that one step leads to another, I let the feet walk their way back into the landscape, joining animal mind to animal body, animal body to animal home.

In the parklike setting of the open savanna, the oaks define the character of the landscape. Coast live oaks give way to drought-tolerant blue oaks on the drier hills. Slimmer and more airy, the blue oaks shimmer a pale bluish green in the afternoon heat. An occasional black oak with dark furrowed trunk and deep green leaves stands out among the blue canopies. The more rounded valley oaks spread graciously among the upright blue and black oaks.

The blue oak savanna has been called the Pacific coast equivalent of the Serengeti Plain. Oak-dotted grassland dominates the foothills rimming the Central Valley, from the south-facing slopes of the Russian River to the open

hillsides of Salinas Valley and up the flanks of the Sierra Nevada. The configuration of scattered trees in a sea of grasses and wildflowers evokes the ancestral landscape of Africa, home of early humans, home of first feet walking. Ten million years these oaks have been in California, long before the first humans walked in Africa.

Walking into this spacious time, my ancestral body enters the ancestral landscape. I relax with the comfort of being able to see in all directions. Capable of perceiving threat, my body is at ease, confident it can protect itself. The ground is firm and easy to walk on, the trees open and companionable. With little danger and few obstructions, the oak savanna seems made for people, a natural fit for survival.

In the warmth of the afternoon sun I wander off the trail into an arboretum of blue and Oregon oaks. With their compact canopies of crooked branches, the blue oaks stand only thirty feet high. In contrast, the Oregon oaks have a classic shape, their unbranched trunks straight, their leaves broadly lobed. Several other oaks have oddly variable leaves, perhaps hybrids of the two.

My feet are looking for the right stopping place. They are ready to be free of the weight they carry. They walk up to a number of oaks, fine and gracious, welcoming and familiar. But none calls loudly, *Come, sit by me.* I wander farther along a narrow deer trail, climbing a small hill in search of a view. *STOP!* Here is the tree, the feet say. Look at this magnificent blue oak! The tree is striking in its girth—an easy eight feet around; the canopy spans over

120 feet. Each of the eight large branches is as wide as the usual blue oak trunk. The tree is old; broken, decaying limbs clutter the ground. All of the growth is at the branch tips, leaving the center of the tree open. The horizontal limbs make curving, dark shadows on the grass. I cannot resist the companionship of this stunning oak.

I lean against the massive trunk and shake the shoes off my feet. To rest in the shelter of this solid and sturdy companion is a treat after walking so many miles. Like others before me I have come to the tree for shelter, for a safe perching spot, a haven for resting. Overhead a cavalry of woodpeckers heralds their territory of acorn caches. *Waka-waka-waka-waka,* the clown-faced birds call, dominating the soundscape, drowning out the chattering of wrens and bushtits. A squirrel jumps from twig to twig. A flycatcher and two warblers chase after insects. The place is raucous with travelers! It is hardly a zone of silence and contemplation.

After moving all day long, I am ready to stop, to simply rest, to do nothing. Though my mind still carries the stories of damaged trees, sole survivors, and complex webs of disagreement over trees, it is time to let even these matters rest. For this I can learn from blue oaks. Blue oaks are masters at resting, especially in the summer dry season. While most oaks are evergreen or deciduous, blue oaks are capable of either state, depending on the amount of rainfall. The leaves use a waxy coating to conserve moisture; this is what gives them their bluish cast. The internal structure of the leaf is reinforced by cellulose and lignin, which grow progressively thicker over the dry season to withstand the physical stresses of water loss. Thus the leaves can lose as much as thirty percent of their water, and the cells will adjust their salt content to prevent wilting. This capacity to survive the cumulative drying heat of the long Mediterranean summer surpasses even desert mesquite and ironwood trees. In extreme droughts blue oaks simply drop their leaves rather than fight the sun. The trees then rest in a state of summer dormancy, quiet until the rains come.

Considering the benefits of rest as a long-term strategy, I drift off into ancestral dream landscapes, wandering among the dozing lions and browsing gazelles. Enough contemplation and investigation, enough ponderous deliberation for now. Let me be an animal, let me find a way with trees that existed before humans. Let me rest on the soft earth soaking up the heat of the day, following the example of the oak, conserving energy, reducing activity.

I awake with a start as a small branch falls to the ground. *Waka-waka-waka,* the rowdy woodpeckers carry on, the party in the air never stopping. I am overcome with a desire to leave the ground and climb up in the tree; all nerve circuitry has reverted to animal mode. Let me be a monkey exploring this tree from the inside out. Enough sitting at the feet of the elders! Let me clamber up the trunk of this inviting oak and look out from its branches like a squirrel or a hawk. Let me follow my ancestral monkey instincts and see what they know about trees.

I bound up from my spot, charged with energy for mounting the massive tree. But the first crotch of the trunk is tantalizingly out of reach. Though I stretch my primate arms as far as I can, this route seems impossible. Perhaps there is another way. I wonder if I can walk up one of the firm, wide branches. This may be easier said than done. But I am so eager to be in this tree, I am willing to try any monkey trick. One helpful branch dips low enough to serve as a first step. The outer portion of the limb is thin and springy, but it holds my weight. This might work, I want it to work. Despite my primate urges I am evolutionarily out of shape, not to mention a bit on the large and gangly side. Oh well, I proceed with unabashed clumsiness. Grasping at nearby branches, I clamber up the first few yards, letting the animal body find its natural locomotion. So far, so good. I'm up ten feet and about a third of the way to the central crotch.

Farther up the trunk, however, I run out of hand- and footholds. In fact there is *nothing* to hold onto except the main branch. Stomach gripping the limb, arms and legs wrapped around the slim branch, I look more like a sloth than an agile monkey. It's laughable! I try the caterpillar mode, slinking along a few inches at a time. But now the drop is over fifteen feet, and the ground looks hard and very far away. Clinging to the branch with less than monkeylike confidence, I find I can't go forward. In fact I'm not even sure I can go backward. In this precarious and unyielding position at twenty feet into the adventure, I am forced to retreat.

How disappointing. My pounding heart and shaky knees remind me of the limiting aspects of mortality. In this extremely serene spot I have succeeded in introducing drama and danger into my life. Isn't this archetypically human? However, I'm determined; there must be another way.

With remarkable vitality my genetic intelligence rises up through the monkey mind. Rocks, tool using—an evolutionary flashback. The protohuman laboriously hauls several flat rocks over to the tree and piles them creatively at the base of the trunk on the uphill side. Yes, this will work. I balance on the stack of stones, stretching easily now to the big crotch in the trunk. I am so pleased with the ingenuity of my biological inheritance. Pulling myself up with primate arms, I grab the main trunk, throw one leg over the gap, and I am home free. Ah, the triumph of reasoning! I am finally here in the arms of this wonderful tree.

Climbing up into this blue oak, all the familiar feelings of being in a tree come alive—how delightful to cavort, to ramble, to explore, to test out every branch and crotch for comfortable resting spots. I scramble up the large side branches, checking for good views, convenient handholds, and other amenities of tree life. The ancestral animal is satisfied; this oak is a superb choice for home, shelter, and resting place. But almost immediately a memory surfaces; I remember my old friend in Santa Cruz, the splendid live oak on the bluff. I didn't know how much I needed a new friend to help me heal from the loss of my old friend. Climbing into the arms of this oak is a welcome homecoming; it gives me a sense of story completed, of tension resolved.

Resting in this oak, I remember the large backyard apple tree I knew as a six-year-old in Buffalo, New York. My second-story bedroom window looked out on this mature tree whose branches almost touched the small white deck off the back of the house. I knew the tree's branching pathways and intoxicating smells intimately. Buried in the thicket of its leaves and fruit, I would play for hours by myself or with a friend, absorbed in the child-sized world of the tree's interior. I was completely at home off the ground, fulfilled in my arboreal habitat.

One day when I was playing quietly on the deck, I gazed back and forth between the tree's branches and the lacy shadows on the porch. For several hours I was mesmerized by the moving patterns of light and shadow. At that moment I understood something important about the passage of time and the changing perspective of the sun on the tree, though I could not put it into words. What I knew then was the power of relationship to Other, based naturally on the grace and truth of spending time in a tree.

Sitting in the large blue oak now, my back against the main trunk, my feet propped up on a branch, I am once again at rest and at home. As the light shifts overhead, I catch the echo of my first connection with trees. I am enfolded by the tree, embraced, upheld. I can feel the healing power in trees that I knew instinctively as a child.

Richard St. Barbe Baker, the great tree-planting saint of England, used to spend at least ten minutes each day with his hands on the trunk of a tree. He said this recharged his energy by connecting him with the tree's powerful circuitry. He was quite serious about this; he recommended it as a natural cure for malaise, stress, and other degenerations of the body and mind. I suspect he knew that the root of the word *druid*, or tree lover, was the same as for the word *truth*. A tree is its own truth, purifying others with this truth as they enter the zone of the tree's energy.

Being in a tree is not the same as sitting beside a tree or underneath one. By entering the space within the tree, I meet the tree on its own terms, subject to the conditions of its shape and history. I fit my body to the tree's body. I enter the web of tree relationships and slow down into tree time. No thoughts, no ideas, no direction, no anticipation. Just the breeze blowing over our shoulders, the silverfish scuttling over our bodies, the sun moving along minute by minute cleansing the ragged mind.

The healing power comes from just being present with the tree. It magnifies the pull to truth, to wholeness. As I am held by a living being, I participate in this powerful force. I enter into the tree's healing presence and merge with the powerful truth of this magnificent blue oak. I rest in the arms of the ancestral tree, at home with my animal body, at ease with our time together.

CHAPTER 21

Gift beyond Measure

ON MARCH 24, 1991, THE FOURTH TALLEST TREE in the world fell down. This 362-foot coast redwood landed with a resounding crash in the middle of a heavy spring storm in the Humboldt redwoods of northern California. A calamity of the dark night, the fallen tree was not discovered until the next day. Apparently the pounding rain had loosened the soil and gusty winds had pushed hard on the old giant. A week before, one of the other large redwoods in the grove had fallen over, knocking into a second tree, like giant pickup sticks in the forest. The second tree fell a few days later, crashing against the champion giant with a weight of hundreds of tons of wood and life force. Already loosened from its moorings and now shaken to its roots, the great tree went down in a heroic plunge to the forest floor, taking a fourth tree with it and scarring several others in the process. For all its extensive history in surviving fire, flood, rot, and insects, the tall tree could not stand against the slamming force of its neighbors.

The famous tree, the Dyerville Giant, was a main attraction in the Founders Grove, one of the first groves to be protected in Humboldt County. Most trees in the grove are at least five hundred years old, and some over a thousand. South of this grove a cross-section of an even older tree counts out at twenty-two hundred growth rings. The Dyerville Giant was younger, but still it was at least thirteen hundred to fifteen hundred years old. It had lived through the Middle Ages, the life of Michelangelo, the invention of airplanes, and the scourge of two world wars. In 1972 the tree won the title of largest tree in the park and was named by the American Forestry Association as its "champion coast redwood." The trunk was fifty-two feet around at the base, and the crown spread seventy-four feet across. The tree was as tall as a twenty-story building and as wide as an average house.

I had heard the news of its demise from a friend in the tree grapevine who urged me to go see it. Drawn by a combination of curiosity and humility, I made the four-hour drive up Route 101 from Muir Beach to Dyerville. The visitor center had a scrapbook of clippings on the tree, and the volunteer hosts were eager to tell its story to a devoted tree pilgrim. At the Founders Grove the brochure said nothing about the tree's change in stature. The news was still too fresh. Visitors would have to cope with the

unexpected on their own. Once on the trail, I passed by the numbered stops for sword fern, redwood sorrel, and tree burls; I was interested only in the Giant. I wanted to see it in its fallen state, its hundreds of years of history now horizontal on the ground.

I was not disappointed. The upright structure of the grove had been completely altered by the impact of this fall. Under the benign conditions of the flatland grove, dozens of full-sized coast redwoods had grown up within hundreds of feet of each other. The outstretched limbs of their root masses had laced the territory beneath the trees into a basket weave of anchors, tying the trees together in solidarity. In the rich soil there was little need for deep roots; even the Giant's roots only penetrated fifteen feet down.

When the Giant fell, it ripped away from this intricate web of history, exposing its root wad to the shock of fresh air and human visitors. The roots measured thirty-eight feet across; the pit in the ground was eight feet deep, a stunning backdrop for a photograph of person and giant. I took the obligatory shot, shuddering at the might and magnitude of this fine specimen of life. The tree was an even more impressive giant lying down. Perhaps it was a reclining Buddha, beginning the next phase of its enlightenment.

Obviously the Dyerville Giant was no longer the tallest tree in the forest. It was not even the longest tree in the forest in fallen position. When it fell, the top third shattered on impact into massive chunks six to twelve feet long. The slabs fell open as if a cleaver had sliced through the wood, popping precut fence posts out of the heartwood. Tattered, twisting bark dripped from the underbelly of the lower trunk, revealing the soft, fibrous, cinnamon inner bark. Fallen logs pinned under the force of the Giant's fall had flattened like pancakes, the grain of the wood splayed in a rainbow fan. One standing tree caught directly in the line of fall showed a ten-foot scrape where the Giant had plowed through its foot-thick bark. The area still resonated with the impact of the fall, the motion captured on the canvas of the forest.

With my small-scale human mind I found it difficult to grasp the magnitude of either the force or the event. The rangers said it felt like a loss in the family, but how do you compare the death of a sixteen-hundred-year-old with that of a seventy-year-old? A young girl, standing on top of the Giant's root mass, said, "Isn't it sad?" She was barely visible from the far end of the fallen tree. The Giant was longer than a football field.

Among the tall redwoods of the North Coast, many before the Giant had caught the attention of gawking white settlers in the early 1900s. One tree felled in 1943 by the Union Lumber Company in Mendocino was measured at 334 feet with a stump diameter of twenty-one feet six inches. It took a twenty-two-foot power-driven saw to cut the oversized tree down. When the loggers finally counted the rings, the tree was found to be 1,728 years old. The

Captain Elam tree, a few miles distant on Little River, was 208 feet tall and twenty feet wide, large enough to build twenty-two houses.

The early settlers measured many trees in their search for the most impressive giant. With each new discovery of yet another astonishing grove, they could hardly believe that nature could be this generous. The straight timber was ideal for building—light yet strong for its weight, free of knots, and straight-grained. Durable in water and soil, and nearly immune to insects, the wood was reliable for a wide range of uses, from shingles to porch rails, wine casks to wharf pilings. It was beautiful as well, pleasing the eye with its soft, rich color. Such trees had not been seen before by white settlers. Because of their remarkable characteristics, they were destined for fame. The trees were a gift beyond measure.

The tall tales of the redwoods drew many woodcutters to the bounty of the Golden State. The trees were irresistible meccas for curious thrill seekers. But after a few decades of harvest, the gift began to look a little shabby. Like bark beetle tunnels on a fallen tree, logging roads had burrowed into the blanket of green, leaving it shredded and torn. The streams were muddy with jumbled debris, the forest floor crumpled from the impact of traffic. The gift no longer seemed infinite.

The Dyerville Giant was here only because of the generosity of early conservationists, who feared for the total loss of redwoods. To increase public awareness of the gift of redwoods, people gave money to set aside some of the groves of giants that had escaped the cutters' zeal. Bolling Grove, Lansdale Grove, Lady Bird Johnson Grove, Rockefeller Grove—the trees now carried the names of those who had given back. As one of the first areas preserved, the Founders Grove was a testimony to the visionaries who saw trees as something other than useful wood. Standing in the grove surrounded by tall trees, I could feel this vision, this sense that others' efforts had made it possible for me to be here in the presence of these trees, that this was exactly what they'd had in mind.

The park rangers said they lose a few old ones like this every year. They go down like dominoes; it's not that unusual. But this was a famous tree. Its obituary was written up in the local paper. People from close and far away made special trips to see the Founders Tree, now laid low for a full casket viewing. It was a massive tree, a history book of enormous proportions, a redwood personage of great bearing. A close-up view of the huge tree was a gift of awesome dimensions.

In the visitor center scrapbook I found a letter to the editor about the fallen tree from a woman in a nearby town. She knew the Giant well and had shown it to her children and grandchildren many times. It saddened her, she said, to think of the tree just lying there rotting. Wouldn't it be better to recycle it? she asked. Use it for firewood, fence posts, shingles, paper? She wrote, "If the old tree had a soul, I'm sure it would have opted to become useful in some way." In her sadness she thought it would be a good idea to do something useful, as if that might lessen the burden of grief. From her perspective, usefulness was a virtue, a tribute to the immeasurable value of this gift.

In the next week's paper a man from a different town replied, calling on readers to pause in respect for the passing of an Old One. He wanted people to reflect on the place of trees in the human psyche and consider the perspective of the tree's lifetime. "Think of the sunrises and sunsets in sixteen hundred years, the animal comings and goings, the winds, the storms, the lightning, the floods, the fogs, the eclipses, the comets, the wars, the inventions, the thunderclaps," he wrote. He chastised the woman for her use-oriented point of view, calling it mean-spirited. In his view the gift of the fallen redwood was inspiration and spiritual reflection.

The commentary continued like this, the tree serving as platform for the ongoing debate on spirit and matter. Local lumbermill operators calculated the fallen giant's million-dollar value with its rich, clear-grain heartwood. Environmentalists extolled the wilderness virtues of leaving the tree to rot in its own time. The controversy polarized people, as controversies tend to do. I suppose someone might have suggested building a temple out of the fallen tree, which would have used the wood *and* cultivated spiritual integrity. The two views of gift were pitted against each other—one on behalf of sustaining human flesh, the other on behalf of sustaining human spirit. Both considered the tree only from the human perspective. No one mentioned the tree's natural usefulness as future soil and nutrients for ferns, mosses, fungi, huckleberry, and salal. Neither view of gift came close to accounting for the magnitude of what had already been given over evolutionary time.

One has to go back further in time to understand the scale of this gift. Seventy million years ago the western edge of North America was a tropical paradise. In this early Garden of Eden, palms, tree ferns, and large-leaved plants clothed the Pacific coast with a gleaming emerald bounty. Under the benign conditions of dry, frost-free winters and wet, rainy summers, the tropical vegetation flourished, shaping a botanical history of abundance for the California landscape.

In time the low-lying coastal rain forests were subject to change. Geologic activity played havoc with the flatlands, uplifting two long mountain chains from north to south. The big bumps on the landscape complicated the flow of air across the earth's surface, setting up a pattern of unequal rain distribution. The coast range and Sierra Nevada claimed most of the storm clouds, leaving less moisture in the rainshadow zones to the east. Summer rains diminished substantially, and temperatures grew more extreme. Under the drying sun and cooling winters, the balmy days of green glory evaporated. In the temperate climate the mixed-conifer forests and redwood ancestors of the cooler Pacific Northwest moved south. The landscape diversified into warm and cold habitats, supporting a mélange of incoming northern species, drought-resistant southern species, and a few remaining relatives of the early tropical ancestors. Though no longer tropical, the climate was still amiable enough to support a wide diversity of emigrants and the occasional ancient relic.

With more drying and cooling, the moisture-loving redwood family was restricted to a narrower geographic range. Once distributed across North America, Europe, Japan, and northern Siberia, the redwoods found conditions in most places less and less hospitable. Following the first glaciation of the Ice Age they gave up altogether, except in protected places free of ice. Only four species of redwoods remained to tell the tale of ancient days: the dawn redwood in China, the bald "cypress" of southeastern U.S. swamps, the giant sequoia in the Sierra Nevada, and the coast redwood along the fog line of the coast range.

After all the mountain-building activity quieted down, the trees settled into long-term colonization of the landscape, cultivating gift-giving relationships with the soil, clouds, fungi, and fluttering birds. In certain undisturbed locations such as the Founders Grove, the trees made a profession of growing, with apparently little to stop them. Hundreds of years passed in the quiet streamside valleys, and the trees turned into giants out of sheer inertia.

The Dyerville Giant is part of an ancient lineage going all the way back to the tropical Garden of Eden. It represents a history older than human memory, a story of giving beyond what humans can imagine. I see the Dyerville Giant as a great teacher, filled with the truth of its own evolutionary integrity. When a great human teacher dies, his or her followers continue to learn from the teacher's presence. The teacher is still alive in the mind of the student. When a great tree dies, it can still teach the student of trees. Its presence remains a conspicuous part of the landscape for many hundreds of years. To choose to leave the tree intact increases the odds of recognizing the tree as teacher. Then if students like me come looking for the gift, they do not have to search through fencerows, stove ashes, and fallen roofs for the fragmented remains.

CHAPTER 22

Offering of Darkness

BLACK SHAWL WRAPPED AROUND MY SHOULDERS, I am watching the day settle down into night. I sit quietly, preparing the mind and body to walk out into the blackness, to meet the trees without the familiar presence of light. I want to know the other side of illumination; I want to meet the place where I can't see, the dark underside of the day.

As human beings we are instinctively drawn toward the light for key information for survival. The light dances, the light entertains, the light gives us confidence in negotiating the visual world. But the light focuses our visual attention on detail and distinction, on near and far. There are other ways of seeing that are hindered by the light. These modes gather information by rod cells, by peripheral vision, by touch and kinesthetic awareness. These are the skills of the night animal. I want to remember them.

For many people living in an electrified world, darkness is a rare experience. City lights blot out the stars for miles beyond the edge of town. Most inner-city children never see the Big Dipper. Light pollution also frustrates astronomers, who require darkness to see great distances. I wonder sometimes if darkness deprivation contributes to the edgy malaise of modern society.

It has taken a week of walking at night to settle in with the darkness of the wild. Where I walk there are no lights, no promenades, no shopping malls, no flashing television screens. Only trees, grasses, and the great empty sky overhead. The softness and stability of these forms have slowly calmed my mind and perception, freeing them from the faster vibrations of the electrical world. With natural light as my tutor, my eyes are relearning the subtleties of dawn and twilight.

As the sun dips below the horizon, the trees begin to lose their green brightness. Rounded oak canopies and the tall Douglas fir spires merge with the ridge to form a single bumpy shape. I watch the pale evening light drift into darkness. In the saturated glow of evening, the bleached blond grasses have turned golden with the last light. It is time to go walking, time to enter the realm of the night trees.

I turn off the main road and walk up the trail toward the coast live oaks. Under their closed canopy the remaining light is substantially reduced. Pale brown leaves shine against the dark earth, brightening the path between black trunks. I walk slowly under the trees, tracking the sounds of my feet on the brittle leaves.

In this early stage of darkness the trees show some gradation in value. The oak trunks before me are rich black, the distant Douglas firs are tall shadows of gray. A few early stars poke out between the drooping oak branches. Tonight is the new moon, darkest night of the month, when all the stars come out and let themselves be seen. New moon, new beginning. A fresh sky to fall into, a fresh meeting with the far reaches of the dark. This is the special gift of the new moon, a chance to align my activity with the lunar rhythm and let it move forward in the natural cycle.

I am writing in the velvety dark, no flashlight or candle to cast a spell on my eyes and blind my night perception. I don't want to be distracted from being with the black trees. My words spill clumsily across the page, unchecked by the methodical, correcting mind. My hand moves freely with the night thoughts, unhampered by the judging eye.

The light is almost completely gone. I can no longer see the tip of my pencil moving on the page. Watching the slow, gradual change from light to dark softens the uncertainty that comes with the loss of light. It is not so shocking as flipping a light switch and being thrown into the scary black. The darkness is kinder than that. There is time in the slow approach of night to shift over from the eye's dominant cone cells to the light-collecting rods, from focused to peripheral vision. There is time to sharpen the capacity for perceiving movement in the dark.

Crickets sing in the warm air, calling out the stars one by one. Summer constellations appear between the branches of the oaks—Scorpio, Sagittarius, the Summer Triangle. I catch glimpses of Antares, Vega, and Deneb—bright stars marking the earth's seasonal movement. Against the purple blackness the Milky Way is a spray of brilliance. Home galaxy, great ribbon of light in the sky! My creature body can barely grasp the thought of this vastness as home.

In books the silhouettes of trees are always flat and perfect, as if they could be cut out and used for night scenery. But these silhouettes before me have depth and texture. Clusters of leaves droop down off the branches, curving and arching with elegant form. The limbs move in space pushed gently by the night wind. I touch the density and roundness of the black shapes. I am one of them. Our bodies are filled with movement; we sing in the dark night.

Black creeps up from the ground, stealing the last light from the once golden grasses and shiny oak leaves. The trail is barely distinct from the forest floor. The trees are quiet here in the night, standing in place through another cycle. The steadiness is palpable. There is a particular comfort in leaning against a tree in the night. Its solid mass and rooted presence provide reassuring contact. It is not the same as sitting by a tree in the daylight, looking out to a distant view. At night the tree becomes a reference point for looking in, a place of grounding from which to wander into the night sky.

I leave the oaks to follow the trail down the hill into the Douglas fir forest. Along this steep grade I walk more slowly and carefully, each step feeling for the shape of the ground. The fir canopy is tighter and denser than the oaks; I cannot see the trail. I can only follow it with my feet. Even my peripheral vision is greatly reduced under the thick branches. My feet choose their steps now by texture and stability rather than by visual clues. I walk with t'ai-chi feet, planting myself in the ground, asking the feet and ankles to dance with the night. Each step, a step into the unknown. Each step, embracing whatever falls along the path. Each step, paying close attention, though I can hardly see anything.

I enter a last stretch of pitch-black on the dark north side of the hill. The slope steepens; I bend my knees and hips more, sinking down into my feet, relaxing the small of my back. Let go of the place that holds, let go of the place that flinches, let go of the place that controls, let go of the place that fears. Just let the ground support me. Listen, the wind is breathing in the trees. I hold my hand in front of me. It completely disappears. And yet I know it is there. The road bends to the right; I step off the solid, compacted dirt to the uncharted forest floor. My feet follow the edge of soft and hard, seeking out the trail through the dark tree tunnel.

Walking in the dark night is a way to practice faith, a way to build my confidence in the unknown. This faith is based in both what is known and what is unknown. I know how to walk forward; the motion is still the same in the dark. But by walking more slowly and carefully, my body makes room for what is not known. Each step is a small act of courage, a chance to practice with uncertainty. In walking into the blackness I learn the feeling of caution. I walk with the limits of what I can't see, guiding me, informing my steps. I see now that any picture of trees is not complete with only what is known. By including what is *not* known, my perceptions and actions are altered; I learn to practice courage in the vastness of what I can't see.

I arrive at the stopping spot and crawl into my sleeping bag. Absorbed in night magic, I find a comfortable spot and settle into the soft earth under a large oak tree. I gaze into the richness of the dark sky and find my place among the branches. The great black arms of the oak form an arching canopy for my bed, a kind shelter for my tender heart. I receive the oak's great spaciousness, letting it fill me up in the dark. I rest in the grace of what cannot be seen yet, trusting that in time it will be revealed. This is enough. I am learning the practice of patience: the art of being present with the dynamics of each moment as it unfolds. The stable tree with its outstretched limbs is a great companion in the study of subtle change. I sleep with this, asking the great oak to accept me in its arms for this night of kindness.

I wake up in the first pale light of dawn, not even dawn. Stars are still out; the sky barely blushes with a hint of color. Hills, trees, grasses stand quietly in silhouette. Fresh smells of early morning arrive clean from the night wind, moist with dew. Presents from the night, presents of spaciousness, presents of the open sky. All these are gifts of the ripening of darkness into day.

Purple sky silence fills the wide branches of the oak. A light breeze carries the first song of the morning into the trees. This is the simple gift of night—that it brings the day. That each thing ripens into the next, that the unknown reveals what is known, that by going into the darkness I wake up in the light. And in that turning is the direction of the movement. By watching the dark actually turn, one learns to recognize subtle shifts in momentum. This simple gift is given every day, over and over. One has only to be present. And with this comes the trust that things can turn and that one is part of that turning just by being there.

In the lavender sky the early grosbeaks join the robins, the swallows dart out from their roosts. First sweet tastes of the day drift in on the night wings. Pale gold, pale green—the trees come back into color, the fog drifts into the low valleys. Rising out of the darkness, the morning is full of all possibilities, all choices, all intentions. The offering of a fresh day, how shall I honor it? The gifts of the dark night—how shall I share them? Can I begin this day with fresh intention to follow the unknown? Can my life today reflect the grace of walking in the dark?

I rise up out of the darkness into the light and bow to the oak. With gratitude I accept the generosity of the night. With gratitude I offer my clear heart to the day.

PART FIVE

Choosing to Act

CHAPTER 23

Arbor Day

A SPADE, A FORK, AND A GARDEN CART — POISED IN readiness for the afternoon tree planting. The warm winter light casts a glow on the tools that will soon open the ground and make room for tender roots. In front of the tools lies a square yard of baby redwoods—five hundred small trees nestled close together like little children all in bed. Tucked in tightly, side by side, the skinny seedlings rest under a shallow cover of soil, adjusting to the local climate. Though some of the needles have turned a burnished copper from a recent cold snap, the roots look healthy and the stems are straight and sturdy. Next to the redwoods is a smaller bed of Monterey pines and another of scruffy Douglas firs. One thousand tiny trees in this intensive-care nursery—all being watched over and protected in preparation for planting.

Today is Arbor Day, a day of renewal for the land and for those who love this land. Arbor Day in California falls in February or March; any later it would run into the dry season. Newly planted trees fare best when they can be nourished by the winter rains percolating through the moist soil. To wait until April would mean starting the tree off at a disadvantage, for the ground would already be drying out.

This particular Arbor Day is taking place at Green Gulch Farm in Marin County. The baby trees are destined to become part of the coastal landscape along Green Gulch Creek. Some will stay in the flatlands, becoming windbreaks; others will curb erosion on the barren hillsides. The babies will join groves of cypress and cottonwood, planted earlier to protect agricultural soil from offshore winds. Tree planters will also become part of the landscape for a day, clambering among the rolling hills, digging into the earth.

Since 1975 Arbor Day at Green Gulch has been the occasion for celebrating generational time for both trees and people. Poster displays of each year's plantings chronicle the steady efforts to return the forest to the grazed slopes. The annual event has become an important ritual of renewal, accumulating over time the virtues of tradition. Each year several hundred tree lovers join the Green Gulch Zen community to plant their visions of the future.

Redwoods, pines, firs—these foot-long sprouts are a gangly bunch of teachers. As students of trees, seeking hope, inspiration, and a chance to reconnect with the landscape, we expect a lot of these youngsters. The slim stems, which look more like prunings than trees, carry our aspirations for a better world, our weighty desires to do good deeds on behalf of the environment. Can these tender shoots bear such a heavy load?

One of the gardeners is taking the seedlings out of bed now, a few at a time. He handles them carefully, wrapping bundles of twenty-five in burlap baby blankets for transport to the planting sites. He dips each bundle into a bucket of water, giving the babies a last big drink before they join the upright world. Soon they will be in the ground and on their own.

As Forest Service seedlings, these new shoots have been raised for mass replanting of thousands of acres. Selected for strength and resiliency, they have been propagated for maximum rates of survival against the stresses of drought and cold. In clear-cut areas, seedlings like these may be the only signs of life after the logging debris has been bulldozed and burned. Recolonizing a denuded landscape must be a lonely endeavor for these young trees. For decades, fast-growing tree stocks have mitigated the long-term impacts of clear-cutting. In many cases the trees take to the land just fine, growing up into reliable if simplistic tree plantations over several decades. But mass plantings don't always make it; sometimes seedlings fail because soil and microclimate conditions have been altered too dramatically by the clear-cut. A healthy seedling does not necessarily become a healthy forest. Even in the benign setting of Green Gulch Farm, there are no guarantees.

In today's community ritual, however, these particular seedlings will receive a little extra care. We will be slow and deliberate, holding each tree in our hands with love and care, practicing mindfulness as we work. We have the luxury of time in this small-scale effort and can choose to pay attention to what we are doing. We can observe how our hands touch the stems—roughly or gently? how the burlap is wrapped—tightly or loosely? how the fork lifts the seedlings out of the ground—like a hand or a plow? how the trees go into the holes—in a hurry or with caution? The extra margin of care may or may not make a difference in the long run. Large-scale weather patterns will override even the most careful of human actions. Though it is impossible to know the outcome of this experiment in caring, at some level one knows the truth of it. It resonates with the planter. I think, this is how I would like to be handled—with gentleness and attention, with respect for vulnerability. Everyone knows that babies of all kinds are vulnerable; roughhousing doesn't work with tiny things.

The community of tree planters has gathered for ritual instructions. The head gardener explains about the root tips, the size of the hole, the procedure for repacking and watering the soil. The roots are the most vulnerable plant parts, she says. They are far more critical than the needles for establishing a new plant. We must take care not to handle the roots as we place the seedlings in their holes. Tiny, invisible root hairs break off when roughly handled. These delicate strands of tissue do the hard work of poking into the unknown. The older, tough roots in the center will anchor the seedling, while the delicate tips penetrate the soil in search of food. Please take care of the tree's capacity to nourish itself, she reminds us.

I head out with a bucket of seedlings to reclaim a formerly wooded hillside. Our group will plant redwoods and Douglas firs into the soft soil by the upper pond. I try to imagine this valley filled with the straight spires of redwoods. Redwoods of the past, redwoods of the future—I can see neither really, even as I transpose my experience of other redwood groves onto the landscape before me. The trees of this valley were possibly once as striking as the giants at Muir Woods. But many along the creek were cleared by early logging. Then the 1906 earthquake hit, and almost every remaining tree was taken for the massive rebuilding of San Francisco. Big Lagoon, at the mouth of Green Gulch Creek, became a lumber port, an exit route for the trees of the landscape.

It is eerie to walk through some of the old forested stretches on the north-facing slopes. Redwood companion plants—thimbleberry, sword fern, and hazel—still dominate the plant community. Most are probably the original plants growing under the old redwoods. But out in the open, unprotected by forest canopy, the bushes have lost their graceful airiness and have grown dense and scrubby. Subject to the harsh drying winds, they now look more like the surrounding chaparral. Even so, I still feel the ghost forms of the redwoods now long gone.

With this ghost vision in mind and the ancestors calling me, I dig my first hole and carefully place a seedling upright in the dark, crumbly soil. This restoration work is a work of spirit and community as much as biology. It is a chance to work together with others to imagine the possibility of a reforested world. Another hole, another baby tree, another drink of water. I take this hand, take these trees, and plunge their roots deep into the dirt, burrowing through the soft, moist soil. This is how root tips must feel reaching into the dark—cool, in contact, never alone, no knowledge of open space. Only closeness, only the faculty of touch for orientation.

Tree roots are not dependent on the faculty of sight; they would not call their growing "groping" in the dark, as if the limitations of touch produce only stumbling. Touch is the most ancient sense for all biological organisms; all critical information is gained by direct contact—flesh to flesh, root to soil, hand to stem, life to life. This is how roots speak to the world. They know the direction down to the center, down to grab hold, down through the living darkness.

In each act of planting I affirm my intention to stand for trees, to be part of the healing rather than part of the destruction. Each slender seedling offers the opportunity to sort through priorities, to take a small step in a direction that seems to make sense. I try to imagine the possibility that generations of people and trees will come after me, enjoying this landscape as forest again. This tree planting is such a simple and naive thing we do, though we glimpse at some deep level how necessary it is. Even trained forestry experts cannot predict the two-hundred-year outcome of tree-planting efforts. The trees are only a small piece of a complex ecosystem. For a forest to become a mysterious, layered, and interwoven tapestry, it will take many more tree falls, fires, insect invasions, fungal-spore dispersions, and animal nests—the stuff of years, decades, and centuries. It is not possible to know how this experiment will turn out. Yet each seedling is a link to this vision of returning complexity.

Holding these trees, my hands reach into the darkness, the unknown, the source of wisdom. As I touch this soil, I become part of the energy circuit that connects me across generations to the future of this forest-to-be. In this act of intention I breathe and listen with the little trees. Another hole, another tree, another drink of water. In this circle of breathing and working together, we penetrate the darkness, the little ones leading the way, holding our hands, helping us meet the ground of renewal. This is how we plant ourselves into the earth, finding our roots with the guidance of our tree teachers. Those who have come before us return again to seek the darkness and grow. Baby trees, show us the way. We are kneeling at your feet and doing the hard work of reaching into the future.

CHAPTER 24

Grand Dragon Oak

FOR ABOUT A MONTH ONE FALL I WAS OBSESSED with acorns. Whenever I had a few hours, I would cruise the valleys and hills for good seed sources. I scouted by car, by bike, on foot over Diaz Ridge, along Franks Valley Road, up the flank of Mount Tamalpais, and out by the coastal bluffs. I prowled under oak branch overhangs, combing the rough ground for acorn droppings. I plucked the ripe fruits from the tips of tangled twigs. Like a woodrat or woodpecker collecting for the future, I, too, was storing up seeds for the winter. But my mind was not on eating; it was on planting.

The season was dry, as usual for California in the fall. A couple of weekend storms in October and November had knocked the early ripeners to the ground. By the time I came to scavenge, many acorns already bore the telltale holes of weevils. I took to harvesting directly from the trees, where more seeds were still intact. The best trees (from my perspective as a human) were those with low-lying branches at the edge of the canopy, loaded with perfectly ripe acorns at arm-picking height. I could fill my pockets in minutes, greedy with delight, triumphant at such a fine discovery.

My kitchen was cluttered with bowls of acorns. I was absorbed in sorting and counting the seeds I'd collected. I'd throw out those with maggot holes and keep the smooth, unpenetrated remainder. I labeled the bowls according to location, considering the relevance of genetics to planting. Like the man in Jean Giono's *The Man Who Planted Trees,* I felt a calling to acorns. However, I was more like a fanatic than a prophet. I loved feeling the perfect shape of an acorn in my hands. I carried acorns for good luck on plane trips; I gave acorns as presents. I placed an especially sleek, two-inch-long acorn on my altar for inspiration.

On some days I came home empty-handed. But the problem was not a lack of oak trees; coast live oaks were common in the coastal woodlands. The problem was that not all oaks are good acorn producers. This year at least, the young oaks with less-developed canopies and smaller trunk capacity for water were producing few acorns. Seed production in a drought year was perhaps too costly for individuals on a marginal energy budget. I turned my focus to bigger trees, presuming age might be a critical factor. But even here I was often disappointed. For whatever genetic and ecological reasons, the big producers were few and far between.

Two weeks before Thanksgiving I set off with a group of fellow tree enthusiasts on a hike to the top of Coyote Ridge, over the Miwok Trail, and down into the next valley. We were thinking about a five-hundred-year plan for Green Gulch restoration and reforestation, and it seemed only fitting that we go walking together into the context of our thoughts. As we dropped over the north side of Diaz Ridge to the rich, forested hillside below Mount Tamalpais, the contrast between the two valleys was striking. Like before and after shots of human settlement, one was lush, the other barren. Green Gulch Valley had been logged and grazed extensively, while much of Franks Valley had been protected by state and national parks. Seeing this gave us a better sense of what might be possible.

The trail followed the contours of the land, heading back into a moist canyon protected from the wind. The first oak we saw was sprawling across the path along the edge of a massive rock. The tree had apparently sprouted on the rocky outcrop, sending its roots down twenty feet to the ground below. The main trunk was upright but leaned out over the hill supporting two main side branches covered with fruticose lichen. We named the tree the Grand Dragon Oak for its serpentine frame and oracle-like presence. The sinewy dragon branch, at least seven feet in circumference, held itself up like a snake, grabbing hold of the rock with massive secondary roots. The distorted trunk must have fallen over at an early stage and recovered well from the fall, not only surviving but sending out two more main branches to form the other side of the canopy. Perhaps at one time the tree was supported by the rock, but along the way a chunk fell out from underneath. The tree had literally pulled the stone apart with the help of gravity and the sheer force of growth.

As soon as I saw the tree, I knew we'd hit gold. Not only was it enormous, but there were acorns on all the branch tips within easy reach. This coast live oak was obviously a heavy producer, because all around its perimeter were young saplings three feet high, marking the zone of seed drop. My acorn fanaticism quickly infected the whole group. The abundance and accessibility of such a fine harvest had us laughing and cavorting with delight, like gleeful children discovering a luscious patch of blackberries.

There was something quintessential about this round of seed harvesting. For these few moments we were a tribe of coast people, collecting our winter's food, sharing our joy and common purpose. Certainly others before us had done this hundreds of times, this dance with the land, the trees, and the seasons. The activity was traditional, only the players now had changed. The local Coast Miwok, like most California Indians, had relied on acorns as the staple starch in their diet. Territorial claims were often based on the presence of specific acorn-producing trees. Acorns were a primary trade item across the state. Surely the local tribes knew about this magnificent tree. I began to see how helpful it would be to know exactly which trees were the best grocery stores. To know this meant a tribe had a working relationship with the whole area. To return to one of these generous trees would be a natural cause for rejoicing.

A couple of friends and I stayed long after the others moved on. We were caught in a spell of time, a déjà vu that plunged us into the rhythm of people and land that had existed for thousands of years. Here in this stream of ancestral experience was a vast store of information and knowledge about oak trees and the landscape. By some archetypal synergy between tree and tribal group, we had walked through an imaginary barrier into a place of recollection that riveted us to these gathering grounds. The strong connection to others in the past was inspiring, but it left us with questions. How do we find our own connection with these trees today? How do we become native in this place where we live?

The day after Thanksgiving we planted the acorns we'd collected in the first celebrational Day for the Oaks at Green Gulch Farm. Our efforts were part of a statewide revival, focusing public attention on declining oak populations and habitats. Redwoods and other commercially valuable softwoods had held the conservation limelight for almost a century; now, before it was too late, the oaks needed attention too. Between 1950 and 1980 over four and a half million acres of California oak forest and savanna had been converted to urban and agricultural use. Though oaks are the dominant trees in over thirty diverse California plant communities, the twenty oak species had been more or less ignored by the forestry profession. Increasing encroachment on oak territory stimulated a new wave of scientific and conservation concern for oak communities. As oak woodland dwellers we wanted to do our part.

The day was overcast, cool, and drizzly—miserable conditions from a human point of view, but quite optimal for an acorn. The rain would moisten the ground and make it easier for the seed to absorb water, swell, and crack, sending out its first growing root tip. We began our preparations by giving our collected beauties one last plunge in a bucket of water. The rotten and infested seeds floated; the solid, unpenetrated ones sank to the bottom. Our restoration site was a small hill above the place where horticulturist Alan Chadwick's ashes were buried. These seeds might someday provide shelter for the remains of this erudite and eclectic man who transformed the art of gardening and inspired the Green Gulch organic farm and garden.

Clumsy with yellow slickers and rain pants, our dedicated band of a dozen volunteers clambered up the hill, shovels in hand. The garden crew had prepared chicken-wire cages to safeguard the young seedlings from grazing. At each site we buried four acorns under several inches of soil, surrounded by a milk carton to ward off nibbling rodents. We planted the acorns directly into the ground to allow them room to grow a deep taproot in their first few years. Around the cartons we staked the three-foot-high cages, closing up the tops to deter browsing deer. It was laborious work in the smeary rain, but the thought of a new oak grove on the hillside in fifty years kept us going.

Once the seeds were in the ground, there was not much to do except wait. The winter rains would come or not, and there might or might not be enough water for the acorns to sprout. During the time of long, dark nights and fires in the fireplace, I thought about these trees-to-be and the threats they faced. Many acorns never even penetrate the soil; they are grabbed up by acorn woodpeckers or gobbled by deer or gophers. In some parts of California, feral pigs consume tremendous quantities of acorns. Even cows can eat up to eighteen hundred acorns a day. Exposed seeds lose moisture rapidly; buried seeds are twice as likely to grow into seedlings. Scrub jays are often quite helpful in planting acorns. With their scatter-hoarding behavior, a single jay can bury up to five thousand acorns in a season, only to find less than half of them later on.

But the acorns we planted were safely buried in the ground. If they sprout, they still face the trials and tribulations of life as tender seedlings. Grasshoppers, field mice, and gophers may find the young oak leaves tasty and nutritious. If a seedling makes it past the six-inch nibbling stage, it may be at the mercy of browsing deer. Young oaks assume strange shapes when pruned into shrubs by constant foraging. Brush fires are also a danger for trees close to the ground. For the first twenty or thirty years oak survival is hardly guaranteed. The odds, in fact, are stacked in favor of those who nibble and hoard acorns for food.

By late spring the following season I was curious to see how the acorns had fared. There had not been much rain, but at least the few storms in March had filled the creek and soaked the ground surface. I climbed the hill through a sea of grasses. Peering among the thick stems, I searched for cages and pink flagging tape. To my delight I saw that most of the hundred or so planting sites had at least one seedling with several hardy green leaves. For the few inches of green I saw above the surface, I knew there were three or four feet of taproot running vertically into the ground, seeking out water.

I felt a strong desire to return to the Grand Dragon Oak to pay tribute and offer thanks. This oak, among others, had given us far more than a few acorns. The activity of planting trees had been a joyful return to the land, but it was only a beginning. I caught a small glimpse of what might be required to return wholeheartedly to this landscape. It is not a matter of an occasional volunteer day or a seasonal hike across the hills. It is a full-time endeavor, an entire shift of priorities. The thought is sobering and challenging. It is not easy to reconsider every aspect of how I relate to the land where I live. It will take some time—month after month, year after year—to learn new ways that make sense. My passion for the work is only just beginner's mind energy. What will it take to sustain this over time?

I pulled the seeds out of my shoes and headed over the ridge toward the Grand Dragon. I followed the trail through the grassland as before, knowing the tree would appear at the edge of the canyon. On this walk I was alone, glad for the solitude and the chance to collect my thoughts. I could not fully grasp the importance of these sprouted acorns in my life, but their lives were affecting mine, as co-dwellers in this landscape. I had begun something here, and I owed it my attention. I thought of the California Maidu song:

> *The acorns come down from heaven*
> *I plant the short acorns in the valley*
> *I plant the long acorns in the valley*
> *I sprout: I, the black acorn, sprout: I sprout.*

I climbed into the central crotch of the tree and looked out from the high spot above the rock to the valley below. Here was the valley of trees, the forested slopes, the example of what was possible. The land softened with the rounded canopies of the trees; my eyes rested in this vision of the future. In the quiet, slow heat of midday I sat in the center of this great mother of trees. Year after year the Grand Dragon had been dropping its wealth as the seasons rolled by. The tree was completely invested in the long view of life; its strategies were well worked out and tested to perpetuate oak trees and oak woodlands. How could I find this long view of the land in my own life?

The acorns were a place to begin, a place to let the beginner's mind open up to the land. For this I offered my thanks. A place to begin is essential. Everything else follows from that.

CHAPTER 25

Traces of a Lifetime

HIGH DESERT COUNTRY — THERE IS VIRTUALLY NO water here. At eighty-five hundred feet I am camped in an arid outpost of the White Mountains. The campground is spartan, empty, rocky. Not many mountain tourists make the journey to this dry landscape. To see this country, one must cross the High Sierra and the Owens Valley and then climb the next mountain range east. It is more than a comfortable day's drive from the coast. In even the most accessible season, conditions are marginal.

This is only the second time I have made the effort to travel this far. I am here to see the bristlecone pines, the world's oldest living trees. I want to ask their blessing; I want their help in remembering the wild source. To spend time in this desert means carrying in food and water, being prepared for harsh sun and wind. It is not necessarily an easy visit. I want to see the ultimate tree elders, the ones who have withstood the test of millennia. A journey to the elders tests the journeyer, keeping the conversation focused on life and death.

It has taken a day to acclimate to the altitude and dryness. Like all deserts the bristlecone country is spare and exposed. The White Mountains lie in the rainshadow of the Sierra Nevada; little is left as the weather moves east to the Whites. Annual precipitation is under twelve inches. There are no comforting, benign, generous conifer forests here. Juniper and piñon pine grow gnarled and stunted, finding only minimal support for their survival.

The bristlecone pines live high on the bright dolomite rock above ninety-five hundred feet. I drive up the road following the spine of the mountains. To the west across the valley the Sierra escarpment rises sharply. The Methuselah Trail begins at the visitor center. The place was opened a few years after George Schulman first discovered the trees' great age. His was a classic story of amazing science—who would have believed a tree could live for four thousand years?

On this visit to the bristlecones I am looking forward to an unusual drama—a midday solar eclipse. People have flown to Hawaii and Australia for this experience of a lifetime. Mystical or scientific, it will be something radically different from the ordinary. In California the eclipse will reach only fifty percent, but I want to watch it from start to finish, tasting the rareness of global commonality. I believe I will feel some sense of solidarity with watchers in other parts of the world as we all pay attention to the motion of time.

The four-mile loop trail to the Methuselah Grove winds in and out of shaded canyons and exposed ridges. On the barren white rock the bristlecones have out-survived all other trees. Though the soil is thin and lacking several critical elements, these pines prefer it. The dolomite rock reflects up to twenty-five percent of the brilliant mountain light, keeping the soil cool and moist.

Each tree stands stark and alone, sculpted figures formed by a history of challenging conditions. Pummeled by powerful mountain winds, the trees have been regularly blasted by sand, ice, and rock. Without being here in winter I can only guess at the stinging forces let loose on anything standing on the mountain. Under the pressures of the extremely arid environment, the old trees have twisted into spirals, drying along the grain of the wood. Some forms are curled and bent at ninety-degree angles, others are jagged and spiky. One tree after another, I find myself exclaiming at each breathtaking shape. It is impossible to walk at a normal hiking pace because of the staggering examples of beauty on all sides.

The twisted wood has taken on the odd shapes of angel wings and dragon faces from the long, slow drying process. The wood does not decompose; the high desert isn't wet enough for anything to rot. Instead it dries out very slowly. Some fallen trunks have been dated at over seven thousand years. That is *very* old wood. In my mind it seems as old as stone. One of the keys to the bristlecone's longevity is the large amount of resinous pitch deposited in the already compact wood. This acts as a natural preservative, reducing moisture loss and decay. The arid climate is actually conducive to longevity.

I stare at the molded trunks in awe: how can one species produce so much beauty? Where the bark has been stripped away, the sinewy wood has matured to a golden hue. The pitch weaves ribbons of black through the gold. Next to the sharp gold and black, the green boughs bring light and life to the twisted trees. The needles contribute to the tree's endurance, for they last twenty years—far longer than most conifer needles that are replaced every one or two years.

By the time I reach the Methuselah Grove, it is midday. The eclipse is beginning. But I am so dazed by beauty that I find it hard to concentrate on the event. I'm not equipped to look directly at the sun, so I track the motion of the moon on paper, checking every five or ten minutes for a change in the sun's shape. The sky has grown dim, casting an odd eeriness on the white landscape. Though I have not seen a lot of animals, whoever is here is very quiet, reading the light cues as early dusk.

For an hour and a half I sit with the moon and sun in the company of aged trees. While thousands gaze at the fiery corona around the sun, I watch the play of bristlecone shadows on the ground. The craggy, spiky shadows dance slowly—the only visible movement around the contorted, living sculptures. In the presence of four-thousand-year-old trees the eclipse seems insignificant. If I am lucky, I will witness one or two total eclipses in my lifetime; a bristlecone will see one hundred. I find it difficult to compare hours to centuries. My body has no experience with thousands or even hundreds of years. Stretch as I might, I cannot walk easily into a conversation with a bristlecone.

I circle back to the visitor center, a little lost in the time warp of the eclipse. The late afternoon sun has returned to full strength, bathing the landscape in brilliance. I decide to walk the short loop of the Photographer's Trail. Though I have seen one remarkable bristlecone after another, I want more. Something is turning in me about their beauty. It is the extraordinary beauty of death. For bristlecones, death is a relative thing. By the time one of these trees passes the lifetime of most world religions, it is ninety percent sculpture and ten percent living being. As the tree slowly closes down its capacity for life, it moves more and more toward pure beauty and form.

On the rocky shale I stop by an exultant spiral sculpture. Spirit Rising, I call this tree, because of its upward gestures and inspirational form. The limbs twist and rise from the ancient trunk, the shape of the tree still holding its powerful presence. Contemplating this remnant of time, I try to imagine a little of the tree's history.

Perhaps the first few hundred years went by uneventfully. The tree made it through adolescence and as an adult produced hundreds of small round cones. Another half century passed—and a foot of soil eroded from under the roots, leaving them exposed to the sculpting wind and ice. While the tree stood fast, wars were fought, cultures exterminated, and land plundered. The bristlecone lost a little more bark to the wind.

More hundreds of years passed. Thousands of babies arrived, soldiers practiced killing, teachers passed on wisdom, and grandmothers died. Winter storms came and went, testing the tree against the harsh elements. The occasional earthquake shook loose a little rock, and the tree lost more bark and soil. As the wood dried out, it became ridged and ragged with the loss of moisture.

Even as it continued to live, the tree was dying, closing off one channel after another to the outer branches until only one growing tip remained. Twisting and eroding through old age, Spirit Rising prepared for its final form. Slowly the radiance of the golden, gnarled wood outshone the brightness of the living needles, until finally the last channel of water closed up in the trunk's core, and the green needles fell away. Carved into beauty by the elements, the tree froze into the landscape, only its shadow left moving.

For another three or four centuries the great celebration of spires, spirals, and waving wood radiated in the golden light. Sunset after sunset the stunning forms made a brilliant statement against the deep sky—a statement about beauty in death. The tree offered its powerful beauty to the land, gracing the mountain with the exquisite truth of its life.

More time passed—more storms, more erosion, more corrosive blasting by the wind, more wars, more babies, more settlers—and the old tree eroded into gray, twisted shards, leaving only traces of a lifetime, traces of an epoch. What remains is only what came before the tree—the sun, the wind, and the scant traces of moisture. In five or six millennia hundreds of eclipses have come and gone, millions of people have come and gone. An infinity of grass seeds, bird feathers, and angry thoughts have come and gone.

The night sky fills with a spray of bright stars. I walk in the open desert, drinking in the vast space of the mountains. These bristlecones! I am amazed by their capacity to take elemental form and then gracefully, after many years, return to the elements, atom by atom. Though I will not live four thousand years, I consider how my life might be the design of my death. What is it that creates the beauty of age in human beings? What shears away excess to expose the form within? I wonder how to best use this life to prepare for death. Can I welcome age as an ally in beauty? Can I learn to shine through the years like the golden bristlecones? Can I return to the elemental, reflecting it back to others as I grow old?

Early the next morning I climb to eleven thousand feet to visit the Patriarch Grove. I have returned to the highest bristlecones to ask for witness for my own struggles to recognize the elemental. I wander through a stand of hundreds of trees, looking for the right place to speak. Old trees, young trees, crippled trees, twisted trees. The new summer growth is brilliant and vital against the bright sky. A three-foot sapling gleams beside a five-thousand-year-old shard. A jagged, spiky tree leans out from the rock. The mountain rings with the truth of existence.

I stop by two trees, whose twisting spirals fit tightly together. These are the ones who will serve as witness. I do not know what I have to say, but they will help me. In the confusion of modern overstimulation it is not easy to know what is essential, what is radically simple and to the core. I am asking the trees to push me to my growing edge, even as they do this so elegantly themselves. What is my deepest understanding? How do I live my life as a witness to this depth of truth? What will I leave behind for others when I die?

These questions penetrate to the core, stripping away all excess. There is no time to waste. I do not have four thousand years to prepare for death. I must act on these questions now and do the hard work of growing into beauty before I die.

CHAPTER 26

Wind, Rock, and Ice

LEMBERT DOME—THE GRANITE MONOLITH RISES up out of the broad sweep of Tuolumne Meadows. High in the Sierra Nevada the dome lifts with graceful, sloping lines out of the pine forest, making a bold fist on the landscape. I am walking up the spine of the dome, following the path of the glacier that sculpted this shape twenty thousand years ago. The rock is lumpy and smooth. Large, two- to five-inch crystals of feldspar jut out from the polished granite surface like buried jewels in the mountain treasure chest.

The dome is one of many famous granite formations at Yosemite National Park, created by the same geologic and glacial processes that made Half Dome, El Capitan, Royal Arches, and North Dome. Domes like this are rare in the world. Granite is the only kind of rock that forms domes. The granite here is similar to other Yosemite granite in texture and chemical composition, but noticeably distinct with its feldspar crystals. Formed five to ten miles below the earth under tremendous pressure, this granite naturally expands once it rises to the surface. Most rising stone releases internal pressure by cracking along joints. But Yosemite's granites have few internal cracks, so great sheets of rock exfoliate in curved surfaces parallel to the dome. The rounded shape reflects the buoyancy in the rock. As I dance lightly up the dome, I, too, find release from internal pressure in concert with the stone.

Once the trail leaves the forest, it takes me out into the wide open sky. There are no guardrails, no signposts, no trail markers, no clinging shrubs—only the windswept granite thrusting out into open space. There is nothing to break the thousand foot fall from the top to the meadow floor below. I revel in the chance to embrace the whole sky, a 360° view. I can't contain my delight; I scamper, leap, and hurtle my way up to the top of the dome, relaxing into the solid, unbroken mass of rock. Though the slope drops away precipitously in three directions, I follow the central route to the summit, aligning myself with the core of the dome's shape.

Tree line, weather line. I am at the edge of the world, where the wind takes over and the trees give up. At nine thousand feet there is virtually nothing growing on the exposed surface of the rock. I shout with the exultation of exposure, the wind howling through my lungs. The top, the very top! I gaze around the circle of the landscape, forests and mountains in all directions. Cathedral Peak, Mount

Dana, Mount Lyell, the Kuna Crest—they ring Tuolumne Meadows with distant grandeur. Here is the actual long view, the physical experience of boundlessness, the reality of belonging to a larger context.

I hop, run, and dart down the north side of the dome to a steep drop-off sheared flat by the glaciers. This is the summit for rock climbers, who pit themselves against the straight wall of granite. The view from here is dazzling; it is unnerving to stand this close to the edge. So I lie down on my belly, magnetized by the boundary between rock and sky. There is no farther to go; I gaze into limitless space.

Finally it is enough. I am saturated with infinity. I walk back to a small windbreak of trees below the top of the dome. Barely recognizable as trees, they lie gnarled and crooked in the hollow, only slightly out of the wind. I draw into the shelter of their low, tight branches, contracting to creature size for security and comfort. The awesome view is temporarily hidden, breaking the spell of expansiveness, pulling me back into myself and the limits of my parched, windblown, and frail human body.

Like the bristlecones, these trees reflect the raw power of the elements—wind, ice, snow, rock, sun—all of them beating down on the needles and branches, pruning them to ragged shapes. The exposed trunks are burnt orange from the aging and drying process, having lost much of their bark in the pounding storm gales. A few purple cones remain on the trees, but most lie in pieces on the ground, crumbled by the elements or broken apart by Clark's nutcrackers.

I emerge from my cocoon to consider who these trees are. I have never seen them before. There are only a few subalpine species in the Sierra; these trees clearly qualify as *krummholz,* or severely wind-pruned trees. It is as if they are embedded in the rock, bound by the contours of the dome. They do not exhibit the usual tree characteristics, standing upright against the shape of the land. These trees crouch out of the way, limited to the air pocket between the plateaus of granite.

Whitebark pines, they are whitebark pines, timberline trees of the highest altitudes. I wondered if I would ever see these trees. I have read about them in field guides, but this is the first time I've seen them up close. The white bark and needles in bunches of five are key clues. So here you are, whitebarks. This is a long-anticipated meeting. How fortunate to find you here.

Crouching with you out of the wind, I am struck by the powerful limits you face. In this infinite sky where I have tasted the limitless, you are a reminder of constraints. Fierce winter, limited water, poor soil, short growing season—all these define who you are. They are the boundaries of your life-form. Living so close to the edge, you take on elemental form, growing as fully as you can under the limits of subalpine existence.

There is tension in my attraction to the infinite. Given a grand view, one becomes a visionary. The sky calls me out to see beyond the ordinary limits of life in human form. I stand in the wind, on the edge of the rock, stripped by the elements to what I share in common with all I see. The shock of this expansiveness awakens the truth of interpenetrating realities. I am the sky, I am the rock, I am the pine. The wind is in me, in my breath; the stone is in me, in my body; the tree is in me, in my mind. The truth of this is real; I recognize the depth of it.

At the same time I exist in the limits of who I am, distinct from all other beings and forms of matter. The wash of the infinite clarifies the sharp edge around my form, naming me as recognizer. In the same moment I experience formlessness and form—complementary tensions, each changing constantly into the other. I flow into the tree, inhabiting a shared conversation, and at the same time flow back into myself, watching the edge of self change. I restrain myself to come into the presence of the Other, even as I make contact across the distinction of boundary.

The sense of unity with this rock, with these trees, speaks to me. These are my relations—all my relations! But the unity is only part of the truth. The yearning for unity arises from the debilitating experience of fragmentation. The complementary pull is to relationality, to mutuality. And this comes from the experience of wholeness. Only if I am completely and fully a human being can I meet these whitebark pines completely and fully as whitebark pines. And I meet these whitebarks fully by meeting all their relations—the wind and the rock, the ice and the sun. In and out, the dance of limits and limitlessness, the movement of unity and relationality.

I leave the shelter of the pines to return to Tuolumne Valley. I cross the open granite dome more slowly, watching my steps on the sloping rock face. The wide view blows through me; I cling to the steady rock. Following a different route to the west, I find another ledge below the summit. A few trees line the edges of the small space; in the center is a circle of rocks. A medicine wheel, I think. Small chunks of granite form a cross in the wheel, lining up with the four directions. All my relations—the four directions. From these four sources come death and rebirth, joy and wisdom, receiving and letting go. The granite pieces date back to the time when glaciers moved across the rock, scraping and grinding the outcrop to its rounded shape. The rocks are fragments of history, dragged by the ice across the great dome.

A glacier carried the stones for this medicine wheel twenty thousand years ago. The sixty-mile-long Tuolumne Glacier filled this valley, carving its way through the solid granite. Only traces are left now—a few hardy whitebark pines in the crumbs of glacial soil, a few scattered rocks arranged in a healing wheel. But the glacier is still present, the glacier is still the primary shaping force of the high granite domes. Freed of the weight of ice, the rock continues to lift and rise. Elemental forces, elemental contact, elemental relationship. I am walking in the glacier's path. I am circling in the granite wheel. Wind and sun, ice and snow—all my relations before me, all my relations around me, all my relations in me—high on the granite dome.

CHAPTER 27

A Multitude of Voices

ONCE THERE WAS A CLEAR, UNOBSTRUCTED VIEW from the ridge above Port Townsend, Washington, across the strait to Vancouver Island. For a time this view was used for strategic defense. It was a functional view, a necessary view—far more important than a merely scenic view. The trees were clipped purposefully to monitor the potential for danger at sea. The northern conifer forest of Douglas fir, cedar, spruce, and hemlock was pruned so that unpredictable neighbors could be watched across the forty-eighth parallel. To guarantee ownership of place, gun emplacements were constructed in the wilderness.

Perched on the northeasternmost tip of the Olympic Peninsula, Fort Worden was assigned the task of protecting the sovereign borders of the American nation-state. The boundary between Canada and the United States was declared to be halfway across the Strait of Juan de Fuca. Any party crossing that invisible line could be perceived as enemy, particularly if it represented competing territorial interests. The fort was built in the midst of a two-thousand-mile conifer forest that filled the peninsula to the brim, pouring down the steep slopes of the Olympic Mountains to the Quinault and Quillayute Rivers and out to the broad beaches of the Pacific. In the wet interior of the peninsula, the temperate rain forest dripped with moss and ferns, and trees grew to swollen dimensions in the absence of interference.

The officers' quarters for the fort stood on the flat plateau above the harbor. A long row of large white houses faced onto the open green, across from military headquarters. From this platform the men were dispatched to seaworthy patrol boats or ridgetop lookouts. Each concrete bunker held food and bedding provisions, water, and ample ammunition for a crude defense of an immeasurable treasure. The simple box fortifications were partially buried in the hills to minimize their presence while still permitting the protrusion of black cannon barrels. The forest was cut back to enhance the strategic view. This view was maintained throughout World War I to guard against invasion by the Japanese. When the war was over, the concrete bunkers finally gave in to the trees that had been trying to grow back the whole time.

Now the forest has thoroughly infiltrated the barricades in a miscellaneous, unstrategic takeover. Sapling alders and wild rose brambles cluster in the sunlight. Their gay, dappled shadows cover the once barren ground; their leaning forms soften the edges of the eroding concrete bunkers. A small crowd of gangly Douglas firs poke through the broken barricades, filling in the gaps left by the military. To a casual observer this might appear to be an average forest. But actually it is a forest coming back, just beginning to be a forest again. Without a war to hold things up, the trees are growing over everything. Leave a spot bare in the northwest forest and moss will cover it over. Leave the soil undisturbed and the seed bank will sprout into full form. Leave the trees alone and they will obscure functional views, inviting a wild thicket of lush growth to take over.

The trees grow like weeds in a lot left untended. They form a blanket of wildness on the land, springing up in the absence of attention. No one has landscaped this arrangement. It is apparently random, chaotic in pattern and shape. Every fern, every alder, every rose tangle is just finding its own way through the competition for space. Some are doing well; others will never make it past the seedling or sapling stage. This is a forest in transition—a bunch of new ideas trying to gain their footing. It is all quite temporary, but left alone long enough, the blanket will persist.

In a bramble of young firs, alders, and thimbleberry lies a half-buried concrete step on the lip of the hill, a fading foothold in the shifting forest. I perch on the step, looking for signs of the former fort. The step doesn't lead anywhere anymore. Ferns have covered over the path, blocking access to the steep hillside. The forest is rebounding with a ferocious resilience, doing its best to heal the trauma. The bramble of plants has taken back the integrity of the place. Like a single tree with a multitude of voices, the grove of young alders speaks a blessing of peace over the former war zone.

Twenty or thirty years after the war, I imagine this tip of land showed only the rudimentary beginnings of a forest—a few fallen logs, random Douglas fir cones blown in from neighboring trees, and a dusting of countless invisible pollen grains and fern spores. A few ground squirrels and Steller's jays gossiped in the open spaces, dropping madrone berries and rose hips. Underneath the surface of the soil, pale fungal threads digested old logs, preparing the ground for future growth of young tree roots.

The red alders probably moved in first, perhaps after a brush fire that would have cleared any remaining vegetation. Taking well to open areas, the alders thrived in the sunlight and enriched the soil for those to follow. In a complex arrangement of friendliness, the alders grew thickened root nodules to host bacteria capable of processing nitrogen gas into ammonia. The trees then converted this ammonia to amino acids, which formed the protein building blocks for cell growth. The return of the forest began in this cellular act of interpenetration, a marriage of producers at the ground level.

In the heat and exposure of the cleared area, Douglas fir cones cracked open, sometimes with the help of mice and chipmunks. When the winter rains came, the seeds germinated, forming sprays of young needles no more than an inch high. The seedlings grew quickly to five-and ten-foot saplings, soaking up the luxury of unblocked sunlight. From a distance they resembled a neglected lot of Christmas trees too scrappy to sell.

In another fifty years the forest will change shape and dimension, growing from youth to early maturity. I lean against a small tree, envisioning the Douglas firs overtaking the alders, squeezing out the sun below their full branches. The alders will die back, falling onto the forest floor, turning over their stored nitrogen to feed the firs. The fungal associates will weave their way into the conifer roots, crisscrossing the soil with a tapestry of mycelia, spewing millions of powdery spores from their fruiting bodies. Warblers, kestrels, and sharp-shinned hawks will lace the airways with flight, while nuthatches, brown creepers, and woodpeckers spiral the trunks as they forage.

In two hundred years, the time of ten human generations, the largest Douglas firs will be rich producers of strong, fine-grained heartwood, dwarfing the old bunkers of the earlier warfare state. Fallen trees will support dense colonies of bark beetles, ants, and centipedes, and fungi will flourish in the rotting logs. Hemlock seedlings will use the soft soil and moisture stored in crumbling nurse logs to gain an advantage over other young trees. Mosses and liverworts will blanket the ground, soaking up the surface runoff like a decorative sponge.

Three hundred years later all traces of war defense will be completely gone, broken down by the steady drip of winter rain over the landscape. Hemlocks, western red cedars, and Sitka spruce will stand in the understory shade below the remaining Douglas firs. Craggy and broken-topped, uneven with jagged branches, the firs will tower over the forest. By the random events of history these firs will have become an ancient old-growth forest, and the memory of a clearing will have been replaced by a web of complexity.

In the first stages of reclamation all this is possible. The order of simplification is easily replaced by the chaos of complexity. Time and a good seed bank are the main requirements. Living systems have a proclivity for the unruly; time naturally generates relationship. To allow this forest, this multitude of voices, to take the first step toward recovery is to begin the process of healing. This seems like a joyful thing to me. As a witness to this healing I feel the wilderness affirmed; I hear the highly cultivated life force speaking powerfully on the landscape.

The voice of a forest is an elusive thing. It sings in the sweet warbles of purple finches and Swainson's thrushes. It rustles in the leaves dancing in the afternoon sunlight. It buzzes in the slim sounds of crickets and mosquitoes. It creaks in the sway of tree trunks rubbing against each other. I wonder when a tree gains its voice. How old must it be to speak from its position in the forest? Are the young ones part of the multitude or do only the grand sages claim a voice? The conversation of a forest is a babble of energy flow, an explosion of growing, a richness of intelligence in tree form.

What does this forest have to say? Perhaps its story is just the simplicity of being here. Day after day, year after year, branches growing, trunks expanding—just rising up into being. In the convenience of being ignored, this forest is becoming something unplanned and undesigned, unexpected and unnecessary. In simply following the natural order, Douglas firs, red alders, wild rose, and thimbleberry have co-created a testimony to the miracle of regeneration.

Though it happens independently of any human goals and desires, I feel a need to be part of this re-arising. People need this blanket of green wildness to walk into and remember complexity. This spontaneous becoming is reassuring witness to the power of life. I want to feel certain that the forest will come back no matter how hard it is beaten down.

But is it possible that a forest can be too traumatized to recover? This belief in the possibility of resilience may be a beautiful illusion, based on evolutionary memory rather than present reality. Severe burns, erosion, and climate change can easily reduce the chances for restoration. Yet in our minds we depend on this thing we know about trees—that they will grow over everything. But what if the cumulative and synergistic damage of soil, air, and water challenges this resiliency beyond its capacity? Individual trees may grow back, but the system may be crippled. Many forests have suffered extreme damage in the wake of territorial disputes. So easily the human order displaces the wildness to serve the dictates of conflict.

A healthy forest is rich with structural complexity in the form of tangles, cracks, cavities, and diverse surfaces supporting other organisms. A microspot may be a perching ledge for a butterfly, a nest site for a junco, a crawl space for a centipede. A simplified forest has fewer resting spots, fewer safe places for small creatures. A forest retarded in recovery stays simple, restricted from reaching its full capacity to support life.

In the desire to aid a broken forest, people plant trees—sometimes as an investment in the future, sometimes as a hopeful peace offering. Inevitably the human order makes its own statement, designing its own forests. In some cases, despite the deep and heartfelt need for healing, the trees come out in rows, like soldiers lined up on parade. Tree planting is not a wild activity; it will not necessarily generate randomness. And wild is random, not ordered by any human mind or idea of pattern. There are too many forces at work on every seed and sapling to predict the shape of a forest. It is ungraspable by the solitary human mind.

As I sit here in this recovering forest, I know I could never have planned the fabulous display of intricate shadows dancing on these leaves. I could not predict that just at this time of day the light would be so splendid on the trunks of the alders. I'm not sure it's possible to create the multitude of forest voices just by planting a few trees. With even the greatest of efforts, the human order cannot duplicate a forest's complex attributes.

In this small haven on the top of the bluff, the forest's voice is growing back and speaking from the land. It is blessed with neglect and the kindness of the Pacific maritime climate. The memory of war is being obliterated by random acts of beauty, noticed or not. By mutual preference for peace, the people and the forest have accommodated each other's growing habits. The nation-state has been at peace long enough to allow the forest to grow back in some places. How long will it take for it to grow back in our minds?

Notes

These notes acknowledge both the lineage of human minds and the biological lineage of trees and places informing this writing.

INTRODUCTION

2 Several translations are available of Martin Buber's *I and Thou*. One is by Walter Kaufmann (New York: Charles Scribner's Sons, 1970). See especially his section on trees, 57–59.

3 In Japanese, *shikan* means pure, "only for it," *ta* is a very strong action, and *za* means sitting. So *shikantaza* means pure sitting for itself, not for something else, the action of expressing one's whole character.

4 For an overview of California habitats, see Robert Omduff, *Introduction to California Plant Life* (Berkeley: University of California Press, 1974), and Elna Bakker, *An Island Called California* (Berkeley: University of California Press, 1984).

5 The case against television's promotion of false views of the natural world has been constructed by Jerry Mander in *Four Arguments for the Elimination of Television* (New York: Quill, 1978), especially 53–80, 299–336. See also *The Age of Missing Information* by Bill McKibben (New York: Random House, 1992).

For analysis of despair and other psychological responses to the environmental crisis, see Joanna Macy, *Despair and Personal Power in the Nuclear Age* (Philadelphia: New Society Publishers, 1983), and Theodore Roszak, *The Voice of the Earth* (New York: Simon and Schuster, 1992).

7 For examples of tree symbology, see "The Tree of Life," *Parabola* 14(3): 11–47, 64–71.

8 For an introduction to the philosophy of deep ecology, see Bill Devall and George Sessions, *Deep Ecology* (Salt Lake City: Peregrine Smith Books, 1985); Arne Naess, *Ecology, Community, and Lifestyle* (Cambridge: Cambridge University Press, 1989); and Warwick Fox, *Toward a Transpersonal Ecology* (Boston: Shambhala, 1990). The challenge of authentic "being-with" is also addressed in *Alone with Others* by Stephen Batchelor (New York: Grove Press, 1983).

9 The Mountains and Rivers Sutra is from Eihei Dōgen's classic work, *Shobogenzo*, 97–107. One excellent translation is by Kaz Tanahashi, *Moon in a Dewdrop* (San Francisco: North Point Press, 1985).

For further commentary on the Zen understanding of nature, see "Dōgen on Buddha-Nature" by Masao Abe, *Zen and Western Thought* (Honolulu: University of Hawaii Press, 1985), 25–68.

CHAPTER 1: CLOSE TO WATER

15 These sycamores (*Platanus racemosa*) are rooted at Camp Shalom, Malibu, California, on the western slope of the Santa Monica Mountains.

16–18 Natural history information on sycamores is drawn from Stephen F. Arno, *Discovering Sierra Trees* (Yosemite: Yosemite Natural History Association, 1973), and Donald Culross Peattie, *A Natural History of Western Trees* (Boston: Houghton Mifflin, 1950), 491–496.

NOTES

CHAPTER 2: CALLED BY ALDERS

23 Key texts on Buddhist practice by Thich Nhat Hanh include *Being Peace, The Heart of Understanding,* and *Transformation and Healing,* all published by Parallax Press in Berkeley, and *The Miracle of Mindfulness* (Boston: Beacon Press, 1992).

24 The white alders of this conversation (*Alnus rhombifolia*) grow at Jikoji Zen Center near Saratoga Gap in the Santa Cruz Mountains. The retreat center is part of an open space preserve for Los Altos and Palo Alto, protecting the viewshed and watershed from development.

25 For commentary on barriers to perception and interspecies dialogue, see David Abram, "The Perceptual Implications of Gaia," in *Dharma Gaia* (Berkeley: Parallax Press, 1990), 75–92.

CHAPTER 3: RED FIR ENCOUNTER

30–31 The six climatic regions are described in Arno, *Discovering Sierra Trees,* p. iii. Another useful tree field guide is P. Victor Peterson and P. Victor Peterson, Jr., *Native Trees of the Sierra Nevada* (Berkeley: University of California Press, 1975). For a classical account of Sierra trees, see John Muir, *The Yosemite* (Garden City: Doubleday, 1962), 66–110. I have also drawn on information in Peattie, *A Natural History of Western Trees,* 79–86, 201–205.

34 I met these red firs (*Abies magnifica*) on the road to Yosemite Creek campground, four miles off Tioga Pass Road in Yosemite National Park.

CHAPTER 4: FRIENDS OF THE FAMILY

37 This encounter with coast redwoods (*Sequoia sempervirens*) took place in Marshall Meadow off Empire Grade Road above the University of California, Santa Cruz, campus.

38 For an excellent guide to the wildflowers and shrubs of the redwood community, see Mary Beth Cooney-Lazaneo and Kathleen Lyons, *Plants of Big Basin Redwoods State Park* (Missoula, MT: Mountain Press, 1981).

40 Redwood sprouting is described in *Muir Woods Redwood Refuge* by John Hart (San Francisco: Golden Gate National Parks Association, 1991), 8.

CHAPTER 5: MAPLE ECSTASY

48 These maples (*Acer macrophyllum*) grow along Redwood Creek in Muir Woods National Monument, part of Golden Gate National Recreation Area. This park protects one of the last remaining stands of old-growth coast redwoods in the San Francisco Bay area.

49 Natural history information is drawn from Stephen F. Arno and Ramona P. Hammerly, *Northwest Trees* (Seattle: The Mountaineers, 1977), 190–194.

CHAPTER 6: REFERENCE POINT

55 This California bay laurel (*Umbellularia californica*) lives north of Santa Rosa, California, at Pepperwood Nature Reserve of the California Academy of Sciences.

56 For further reflections on walking and place, see Gary Snyder, "Blue Mountains Constantly Walking" in *The Practice of the Wild* (San Francisco: North Point Press, 1990), 97–115.

57–59 The landscape history of Pepperwood Ranch was provided by Greg de Nevers, resident biologist.

CHAPTER 7: MAGNETIC PRESENCE

61 For an introduction to the role of fire in the California landscape, see Bruce M. Pavlik, Pamela C. Muick, Sharon Johnson, and Marjorie Popper, *Oaks of California* (Los Olivos, CA:

Cachuma Press, 1991), 56–57, 124–125. The fire history of the area was provided by Greg de Nevers, resident biologist, Pepperwood Nature Reserve.

62 The large Parry or common manzanitas (*Arctostaphylos manzanita* Parry) and the smaller Eastwood manzanitas (*A. glandulosa* Eastwood) grow above Garrison Canyon at Pepperwood Ranch, Santa Rosa, California. *A. glandulosa* regenerates by stump sprouting; *A. manzanita* regenerates by seed.

CHAPTER 8: LIFETIME LOVERS

71 This tan oak (*Lithocarpus densiflorus*) and madrone (*Arbutus menzeisii*) keep each other company on the north ridge above Philo in Anderson Valley, Mendocino County, California.

73 For a beautiful account of madrones, see Peattie, *A Natural History of Western Trees,* 663–666.

CHAPTER 9: MYSTERY PINE

77–78 For further information on the plundering of Thai and Burmese forests, see updates at Rainforest Action Network, www.ran.org.

81 I later found out these trees are Italian stone pines (*Pinus pinea*). They are planted near Gianinni Hall on the northwest corner of the University of California, Berkeley, campus.

83–84 For further exploration into the mind of wilderness, see Roderick Nash, *Wilderness and the American Mind* (New Haven: Yale University Press, 1973).

CHAPTER 10: THE GOLDEN TIME

85 This *Ginkgo biloba* grows between Gianinni Hall and the north fork of Strawberry Creek on the University of California, Berkeley, campus.

86 Useful evolutionary information on ginkgos can be found in Andreas Feininger, *Trees* (New York: Viking Press, 1968), 30, 60.

CHAPTER 11: A WAY OF LOOKING

91 This elm in the Cedar Hills suburb of Portland, Oregon, has now been replaced by a young Norway maple.

CHAPTER 12: FALLEN TREE

104 This Monterey pine (*Pinus radiata*) lived on Sunset Avenue in Muir Beach, Marin County, California.

105–106 For consideration of coastal pine genetics, see William J. Libby, "Genetic Conservation of Monterey Pine and Coast Redwood," *Fremontia* 18(2):15–21. See also Holmes Rolston III, "Duties to Endangered Species," in *Philosophy Gone Wild* (Buffalo: Prometheus Books, 1989), 206–220.

CHAPTER 13: HOUSE OF WOOD

111 For a poignant account of the staggering loss of the world's forests, see articles in *Biodiversity,* E. O. Wilson, editor (Washington, DC: National Academy Press, 1988).

CHAPTER 14: BONES IN THE LAND

117–119 This open landscape of coast live oaks (*Quercus agrifolia*) lies off Empire Grade, west of the University of California campus in Santa Cruz. For a geologic history of the coastal marine terraces, see Burney LeBoeuf and Stephanie Kaza, eds., *The Natural History of Año Nuevo* (Pacific Grove: Boxwood Press, 1981), 78–85.

126 For more reflections on the process of becoming native, see Van Andruss, Christopher Plant, Judith Plant, and Eleanor Wright, eds., *Home! A Bioregional Reader* (Philadelphia: New Society Publishers, 1990); Gary Snyder, "The Place, the Region, and

the Commons," in *The Practice of the Wild* (San Francisco: North Point Press, 1990), pp. 25–47; and Jim Cheney, "Postmodern Environmental Ethics: Ethics as Bioregional Narrative," *Journal of Environmental Ethics* 11(2):117–134.

CHAPTER 15: OVERTURES OF PEACE

130 These coast redwoods (*Sequoia sempervirens*) dwell along Cave Gulch Creek near Empire Grade Road, west of the University of California, Santa Cruz, campus.

134–135 For a history of human destruction of the redwoods, see Peattie, *A Natural History of Western Trees,* 20–27, and Joseph E. Brown, *Monarchs of the Mist* (Point Reyes, CA: Coastal Parks Association, 1982), 35–39.

CHAPTER 16: LINEAGE OF FEAR

137 This fragment of old-growth Douglas fir forest (*Pseudotsuga menzeisii*) survives in the McDonald Research Forest of the University of Oregon, Corvallis.

144 For a history of human use of old-growth Douglas fir forests and an ecological analysis of timber cutting, see Elliot Norse, *Ancient Forests of the Pacific Northwest* (Washington, DC: Island Press, 1990), 27–32, 271–278. For reflections on the history of fear and the forest, see Catherine Caulfield, "The Ancient Forest," *The New Yorker,* May 14, 1990.

CHAPTER 17: PILGRIMAGE

151 This particular species of fig tree (*Ficus religiosa*) is commonly known as the Bodhi or Bo tree in reference to the Buddha's awakening. The tree is also sacred to the Hindus as the site of Vishnu's birth. The present Bodhi tree at Bodh Gaya is likely a direct descendent of the original tree.

152–153 The giant redwoods (*Sequoiadendron gigantea*) of Calaveras State Park are protected in two groves, the North and South Groves, just off Highway 49. Information about the trees in the groves has been taken from interpretive trail guides written by Wendy Faris.

A.T. Dowd was only one of a number of people to claim to have "discovered" the giant redwoods. For more complete information on this and other aspects of Sierra redwoods, see Richard J. Hartesveldt, H. Thomas Harvey, Howard S. Shellhammer, and Ronald E. Stecker, *The Giant Sequoias of the Sierra Nevada* (Washington, DC: US Department of the Interior, 1975); also Stephen Arno, *Discovering Sierra Trees,* 37–43.

157 For an overview of some of the concerns regarding management of the fire-adapted redwoods, see pp. 146–150 in Hartesveldt, et al., *Giant Sequoias of the Sierra Nevada*; also Howard T. Nichols, "Managing Fire in Sequoia and Kings Canyon National Parks," *Fremontia* 16(4):11–14.

CHAPTER 18: THE ATTENTIVE HEART

159 This retreat took place in Anderson Valley, near the town of Philo, California. The title of this piece is taken from the title of the retreat, led by Joanna Macy and Christopher Titmuss.

162 Use of the mindfulness bell to stabilize the attention has been promoted in the United States by the Vietnamese Zen teacher Thich Nhat Hanh. For examples of bell use in classrooms, meetings, and at home, see "The Mindfulness Bell," a newsletter published by his students.

164–165 The devastating ecological impacts of timber operations are outlined in Norse, *Ancient Forests of the Pacific Northwest,* 161–221.

167 The tension of world work vs. inner work is common in religious practice. For Buddhist commentaries on their mutual

relevance, see Robert Aitken, *The Mind of Clover* (San Francisco: North Point Press, 1984); Fred Eppsteiner, ed., *The Path of Compassion* (Berkeley: Parallax Press, 1988); and Sulak Sivaraksa, *Seeds of Peace* (Berkeley: Parallax Press, 1992).

CHAPTER 19: CUTTING WOOD

169 The practice of koan study is central to the Rinzai school of Zen Buddhism. For an introduction to koan study, see Robert Aitken, *Taking the Path of Zen* (San Francisco: North Point Press, 1982), 95–109, and *The Gateless Barrier: The Wu-Men Kuan (Mumonkan)* (San Francisco: North Point Press, 1990).

CHAPTER 20: HELD BY A LIVING BEING

178 For further information on the California oak savanna, see Pavlik et al., *Oaks of California,* 51–73.

180 The large blue oak (*Quercus douglasii*) grows at Pepperwood Nature Reserve above Martin Creek.

181–182 Drought adaptations of blue oaks are described in Pavlik et al., *Oaks of California,* 53–54.

CHAPTER 21: GIFT BEYOND MEASURE

187 The Dyerville Giant (*Sequoia sempervirens*) can still be seen at the Founders Grove, a mile north of Humboldt State Park headquarters along the Avenue of the Giants, off Route 101.

192 The diverse capacities of redwood for human use are thoroughly described in Peattie, *A Natural History of Western Trees,* pp. 21–23. The history of park protection is outlined in Brown, *Monarchs in the Mist,* 37–41. See also Kramer Adams, *The Redwoods* (New York: Popular Library).

193 Letters to the editor ran in the *Times-Standard,* Eureka, California, April 4–8, 1991.

195–196 The postulated evolution of California's flora is laid out in Peter H. Raven and Daniel I. Axelrod, *Origin and Relationships of the California Flora* (Berkeley: University of California Press, 1978). Changing patterns of redwood distribution are described in Ralph W. Chaney, *Redwoods of the Past* (San Francisco: Save-the-Redwoods League, 1979).

CHAPTER 22: OFFERING OF DARKNESS

199 For information on training peripheral vision for night acuity, see Nelson Zink and Steven Parks, "Nightwalking, Exploring the Dark with Peripheral Vision," *Whole Earth Review* 72, (1991):4–9.

200 The trees of the dark are California coast live oaks (*Quercus agrifolia*) and Douglas firs (*Pseudotsuga menziesii*) of Pepperwood Nature Reserve.

CHAPTER 23: ARBOR DAY

209 Arbor Day is celebrated throughout the United States on variable dates, depending on the best conditions for planting. In most Eastern states, tree-planting activities occur in April or May. For an excellent guide to urban and community tree planting, see Gary Molt and Sara Ebenreck, eds., *Shading Our Cities* (Washington, DC: Island Press, 1989).

CHAPTER 24: GRAND DRAGON OAK

218 *The Man Who Planted Trees* by Jean Giono (Chelsea, VT: Chelsea Green Publishing, 1985) is a fictional story of hope about a devoted tree planter in France who, over time, establishes an entire forest, returning life to a once-barren area.

219 The Grand Dragon Oak is a coast live oak (*Quercus agrifolia*), a dominant oak of California coastal landscapes.

220 For Indian use of acorns, see Kat Anderson, "California Indian Horticulture," *Fremontia* 18(2):7–14.

221 For a history of human oak use and scientific study, see James R. Griffin and Pamela C. Muick, "California Native Oaks: Past and Present," *Fremontia* 18(3):4–12.

223 On seed losses to animals, see Mark Borchert, "From Acorn to Seedling: A Perilous Stage," *Fremontia* 18(3):36–37.

225 The Maidu song appears in Pavlik, et al., *Oaks of California*, 102.

CHAPTER 25: TRACES OF A LIFETIME

227 Bristlecone pines (*Pinus longaeva*) are found only at limited high altitudes in California, Colorado, Arizona, New Mexico, Nevada, and Utah. The White Mountain site is one of the most accessible, above the town of Big Pine on White Mountain Road.

228–230 Natural history information on bristlecones is from Russ and Anne Johnson, *The Ancient Bristlecone Pine Forest* (Bishop, CA: Chalfant Press, 1978).

CHAPTER 26: WIND, ROCK, AND ICE

235 For more information on Yosemite domes and geology, see Mary Hill, *Geology of the Sierra Nevada* (Berkeley: University of California Press, 1975).

238 Natural history information on whitebark pines (*Pinus albicaulis*) is from Arno and Hammerly, *Northwest Trees*, 23–28.

240 For considerations of boundary and identity, see Joanna Macy, "The Greening of the Self," in *World as Lover, World as Self* (Berkeley: Parallax Press, 1991), 183–192.

CHAPTER 27: A MULTITUDE OF VOICES

246–248 For information on patterns of succession in old-growth forests, see Norse, *Ancient Forests of the Pacific Northwest,* 33–56.

249 For questions of resiliency and fragmentation, see Michael E. Soule, ed., *Conservation Biology: The Science of Scarcity and Diversity* (Sunderland, MA: Sinauer Associates, 1986).